THE USBORNE
INTERNET-LINKED
LIBRARY OF SCIENCE
HUMAN
BODY

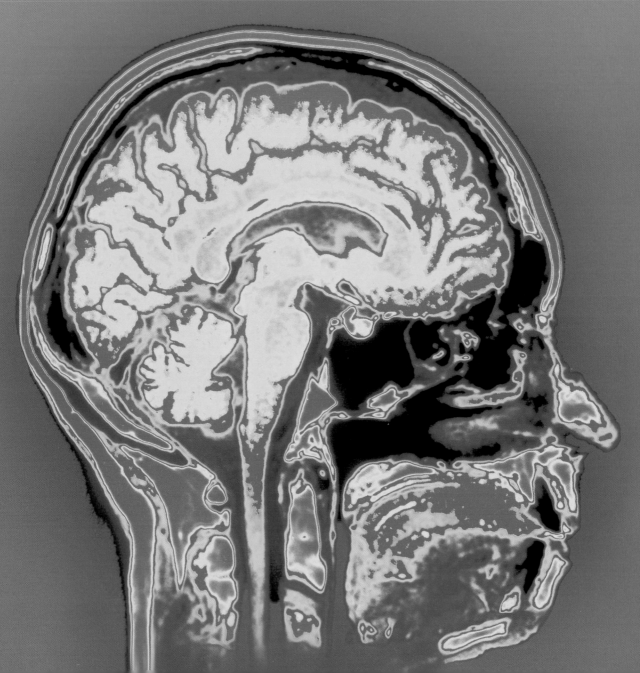

5/14/08, $18

Usborne Publishing has made every effort to ensure that material on the websites recommended in this book is suitable for its intended purpose. All the sites in this book have been selected by Usborne editors as suitable, in their opinion, for children, although no guarantees can be given. Usborne Publishing is not responsible for the accuracy or suitability of the information on any website other than its own. We recommend that young children are supervised while on the Internet, and that children do not use Internet chat rooms.

First published in 2001 by Usborne Publishing Ltd,
Usborne House, 83-85 Saffron Hill, London EC1N 8RT, England.

www.usborne.com

Printed in Spain

AE First published in America, 2002.

THE USBORNE
INTERNET-LINKED
LIBRARY OF SCIENCE
HUMAN BODY

Kirsteen Rogers and Corinne Henderson

Designed by Jane Rigby, Karen Tomlins
and Adam Constantine

Digital illustrations by Verinder Bhachu
Digital imagery by Joanne Kirkby

Edited by Laura Howell

Cover design: Nicola Butler

Consultants: Dr Margaret Rostron and Dr John Rostron

Web site adviser: Lisa Watts
Editorial assistant: Valerie Modd

Managing designer: Ruth Russell
Managing editor: Judy Tatchell

INTERNET LINKS

Throughout this book, we have suggested websites where you can find out more about the human body. Here are some of the things you can do on the websites:

- take a tour of your digestive system
- probe a virtual brain and find out what its different parts control
- follow a baby's growth and development
- investigate genetics by breeding virtual mice
- take a close-up look at different parts of the body

USBORNE QUICKLINKS

To visit the sites in this book, go to the Usborne Quicklinks Website, where you'll find links to take you to all the sites. Just go to **www.usborne-quicklinks.com** and enter the keywords "science body".

The links in Usborne Quicklinks are regularly reviewed and updated, but occasionally you may get a message that a site is unavailable. This might be temporary, so try again later, or even the next day. If any of the sites close down, we will, if possible, replace them with suitable alternatives, so you will always find an up-to-date list of sites in Usborne Quicklinks.

WHAT YOU NEED

Some websites need additional free programs, called plug-ins, to play sounds, or to show videos, animations or 3-D images. A message will appear on your screen if a site needs a particular plug-in. There is usually a button on the site that you can click on to download it. Alternatively, go to **www.usborne-quicklinks.com** and click on "Net Help". There you can find links to download plug-ins.

www.usborne-quicklinks.com

Go to Usborne Quicklinks and enter the keywords "science body" for:

- direct links to all the websites in this book
- free downloadable pictures, which appear throughout this book marked with a ★ symbol

INTERNET SAFETY

When using the Internet, please make sure you follow these guidelines:

- Ask your parent's or guardian's permission before you connect to the Internet.
- If a website asks you to enter your name, address, email address, telephone number or any other personal details, ask permission from an adult before you type anything.
- If you receive an email from someone you don't know, tell an adult and do not reply to the email.
- Never arrange to meet with anyone you have talked to on the Internet.

NOTES FOR PARENTS

The websites described in this book are regularly reviewed and the links in Usborne Quicklinks are updated. However, the content of a website may change at any time and Usborne Publishing is not responsible for the content on any website other than its own. We recommend that children are supervised while on the Internet, that they do not use Internet Chat Rooms, and that you use Internet filtering software to block unsuitable material. Please ensure that your children read and follow the safety guidelines printed above. For more information, see the "Net Help" area on the Usborne Quicklinks Website.

DOWNLOADABLE PICTURES

Pictures in this book marked with a ★ symbol may be downloaded from Usborne Quicklinks for your own personal use, for example, to illustrate a homework report or project. The pictures are the copyright of Usborne Publishing and may not be used for any commercial or profit-related purpose.

SEE FOR YOURSELF

The *See for yourself* boxes in this book contain experiments, activities or observations which we have tested. Some recommended websites also contain experiments, but we have not tested all of these. This book will be used by readers of different ages and abilities, so it is important that you do not tackle an experiment on your own, either from the book or the Web, that involves equipment that you do not normally use, such as a kitchen knife or stove. Instead, ask an adult to help you.

CONTENTS

THE HUMAN BODY

The human body is like a fascinating and complex machine. It is made up of billions of tiny structures called cells, grouped together to form tissue, which in turn forms organs such as the heart. Most activities that take place in the body are controlled by the brain and nervous system. In this book, you can find out about the many different parts of the human body, and how it can be kept working at its best.

These are human skin cells, seen under a powerful microscope. Biologists examine cells by adding colored stains to them, so their different parts become visible. The purple blob in each one is the nucleus, the cell's control center.

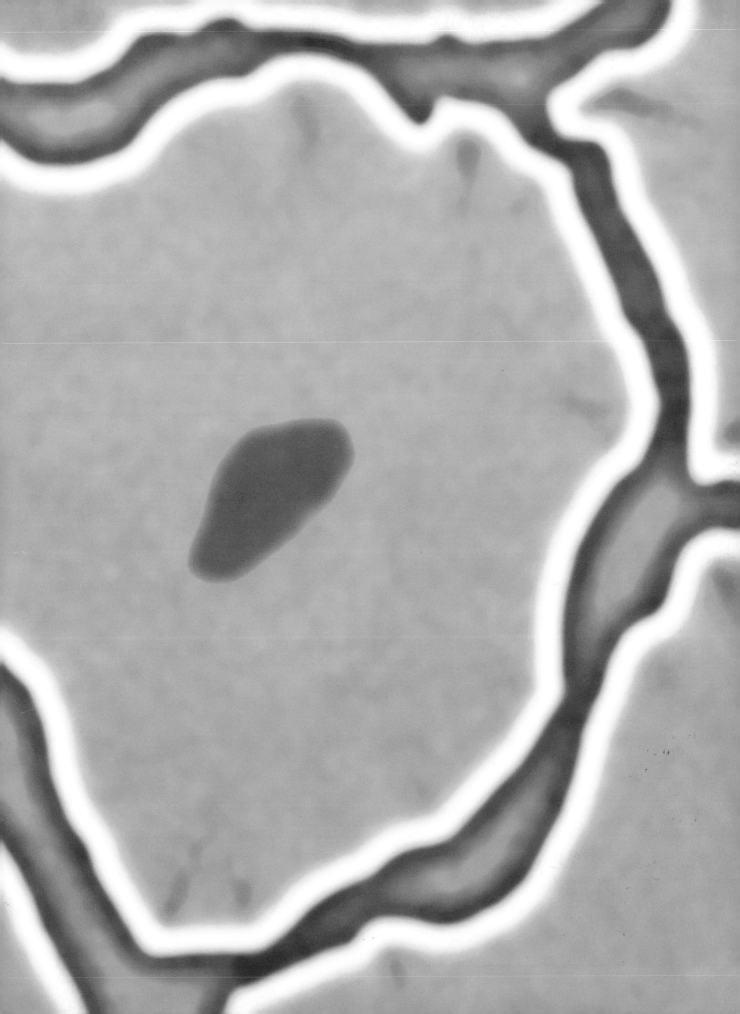

CELLS

Every living thing is made up of one or more tiny units called **cells**. All the processes needed for life, such as producing energy from food and removing waste, take place inside cells.

Cluster of cells, shown at many thousand times its real size

PARTS OF A CELL

There are many different kinds of cells, each with a particular job to do, but most share certain features.

Cells contain a number of small parts called **organelles**, which have various functions. The largest and most important organelle is the **nucleus**. This controls everything that happens inside the cell. It has a double-layered outer skin, called the **nuclear membrane**, and a gel-like middle.

All cells are surrounded by a protective layer called the **cell membrane**, which holds together the contents of the cell. This layer is semipermeable, which means that it lets some substances pass through it, but not others.

The rest of the cell is called the **cytoplasm**. The cell membrane, nucleus and cytoplasm are collectively called the **protoplasm**.

Centrioles play a part in cell division.

Ribosomes help to build up substances called **proteins**, which are needed for all functions within the cell.

Lysosomes can destroy invading bacteria* and parts of the cell which are no longer needed.

Organelles in a typical animal cell (not shown to scale)

The **Golgi complex** stores and distributes the substances made inside the cell.

The **nucleolus** makes the ingredients of ribosomes.

Nucleus. The nuclear membrane has channels called **nuclear pores**, which can open and close to let substances in and out of the nucleus.

Mitochondria convert simple substances into energy for the cell.

Vacuoles are small, temporary sacs in the cytoplasm. They are used as storage areas for liquids or fats.

The **endoplasmic reticulum** is a series of channels used to transport materials around the cell.

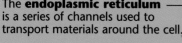

*Bacteria, 48.

8

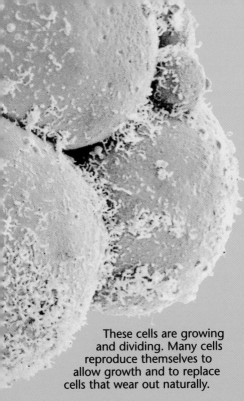

These cells are growing and dividing. Many cells reproduce themselves to allow growth and to replace cells that wear out naturally.

CELL DIVISION

Cells are constantly dying or wearing out, so new ones need to be made. Cells make copies of themselves by splitting into two identical cells, called **daughter cells**.

Stages of cell divison

This single cell is about to start dividing.

The nuclear membrane disappears and the contents of the nucleus begin to pull apart.

The contents reform as two identical nuclei.

A **cleavage furrow** forms, cutting through the middle of the cell.

The cleavage furrow cuts through the cell. Two daughter cells are formed. ★

BUILDING WITH CELLS

Different cell types have different functions. This is called **specialization**. Cells come in a variety of shapes and sizes, depending on their job.

Cells of the same type combine to form **tissue**. For example, **columnar epithelial cells** are long and column-shaped, and allow substances to pass through them. They group together to make a tissue called **epithelium**. This is ideal for lining organs such as intestines, because gases and liquids can pass through it easily.

Columnar epithelial cells

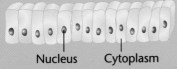

Nucleus   Cytoplasm

Several different types of tissue together form an **organ**, such as the stomach or intestines.

Epithelial cells   Muscle cells

 Cells group together

Epithelial tissue   Muscle tissue

Tissues combine to form the wall of the intestine.

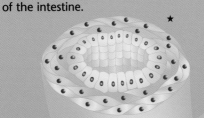
★

SYSTEMS

A group of organs which does a particular job is known as a **system**. For example, the digestive system breaks down food into simpler substances. The human digestive system (see also pages 18-19) contains four main organs: the stomach, liver, pancreas and intestines.

Organs in the digestive system

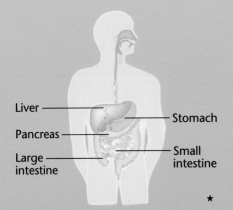

Liver

Pancreas

Large intestine

Stomach

Small intestine
★

Humans also have other systems, including a skeletal system, which supports the body (see pages 10-11), and a circulatory system, which transports blood around it (see pages 14-15). All the systems together make up a living individual called an **organism**.

Internet links

Go to **www.usborne-quicklinks.com** for links to the following websites:

Website 1 Watch a movie about cells.

Website 2 Online diagrams and activities about cells and what they do.

Website 3 A great source of information about everything to do with cells.

Website 4 Amazing electron microscope views of cells.

Website 5 Lots of animal cell diagrams to print out and label.

Website 6 Lots about cells, including an animation showing cell division and a cell diagram and puzzles to print out.

THE SKELETON

Your **skeleton** is a framework of bones that supports your body and gives it shape. It protects delicate parts, such as the heart, and provides hard surfaces for muscles to pull on so you can move.

TYPES OF BONE

The bones in your body can be divided into four main types, depending on their shape.

Flat bones (such as your shoulder blades and ribs) give protection and provide surfaces to which muscles can attach.

Ribs

Short bones are knobby nugget shapes, which are nearly equal in length and width. The bones in your wrists and ankles are short bones.

Wrist bones

Irregular bones have complicated shapes and do not fit into any of the other groups. The bones that make up your backbone are irregular bones.

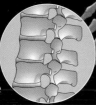
Backbone

Long bones are longer than they are wide. They are slightly curved to make them stronger. The bones in your fingers are long bones.

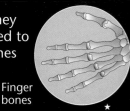
Finger bones
★

DIVISIONS OF THE SKELETON

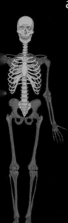

A skeleton can be divided into two parts. The **axial skeleton** (shown in yellow) is made up of bones in the skull, backbone and rib cage. They all lie on or around an imaginary line down the middle of the body. The **appendicular skeleton** (shown in red) is made up of bones on either side of this line, namely those in the arms, legs, shoulders and pelvis.

This picture shows the main bones of the skeleton.

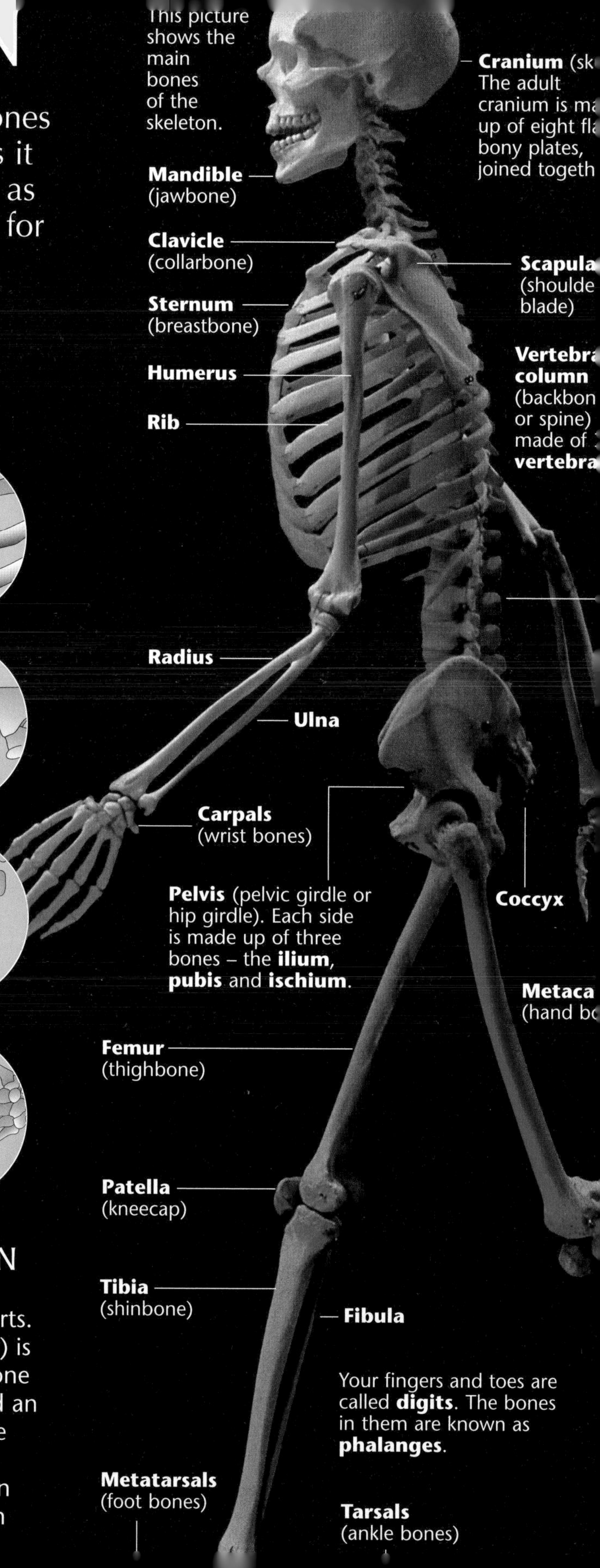

Mandible (jawbone)

Clavicle (collarbone)

Sternum (breastbone)

Humerus

Rib

Radius

Ulna

Carpals (wrist bones)

Pelvis (pelvic girdle or hip girdle). Each side is made up of three bones – the **ilium**, **pubis** and **ischium**.

Femur (thighbone)

Patella (kneecap)

Tibia (shinbone)

Metatarsals (foot bones)

Cranium (sk The adult cranium is ma up of eight fla bony plates, joined togeth

Scapula (shoulde blade)

Vertebra column (backbon or spine) made of vertebra

Coccyx

**Metaca (hand bc

Fibula

Your fingers and toes are called **digits**. The bones in them are known as **phalanges**.

Tarsals (ankle bones)

Joints are places where bones meet. Some, such as those between the bones in your skull, are fixed, but most are movable. The most common types of freely movable joints are listed below. They are called **synovial joints** because they contain a lubricating liquid called **synovial fluid**.

Your hip joint is a **ball and socket joint**. It has a round-ended bone which fits into a fixed, cup-like socket. This lets you swivel your leg in many directions.

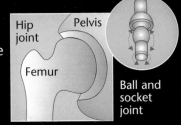

Hip joint — Pelvis — Femur — Ball and socket joint

Your knee joint works like a hinge, so you can bend your leg in two opposite directions, such as up and down. This type of joint is called a **hinge joint**.

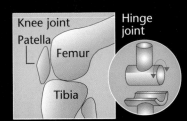

Knee joint — Patella — Femur — Tibia — Hinge joint

The joints in your wrist are **gliding** or **sliding joints**. The surfaces which touch are flat and the bones can move from side to side, and back and forth.

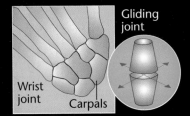

Wrist joint — Carpals — Gliding joint

The **pivot joint** between your top two vertebrae allows you to turn your head from side to side. The rounded end of one bone twists around in a hole in the other.

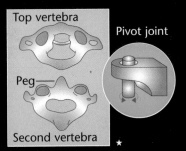

Top vertebra — Peg — Second vertebra — Pivot joint

BABY SKELETONS

The skeleton of a newborn baby has over 300 parts. Most are not made of bone, but of a tough, flexible material called **cartilage**. Over time, this slowly turns into bone, in a process called **ossification**. As the baby grows, some of the bones join together to make bigger bones. By the time it is an adult, its skeleton contains only 206 separate bones.

Each of your bones is covered by a thin layer of tissue called **periosteum**, which contains cells for growth and repair. Inside this, the bone itself is made up of blood vessels, nerves and living bone cells called **osteocytes**, all held within a framework of hard, non-living material containing calcium and phosphorus.

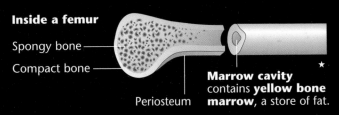

Inside a femur

Spongy bone — Compact bone — Periosteum — **Marrow cavity** contains **yellow bone marrow**, a store of fat.

Spongy bone is a mesh of branches called **trabeculae**, with large spaces between them. This strong, light tissue is found in short, flat bones, and in the ends of long bones.

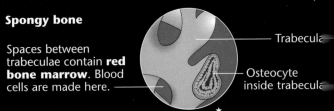

Spongy bone

Spaces between trabeculae contain **red bone marrow**. Blood cells are made here. — Trabecula — Osteocyte inside trabecula

Compact bone is made up of dense, circular layers of bone called **lamellae**. Compact bone forms the outer layer of all bones.

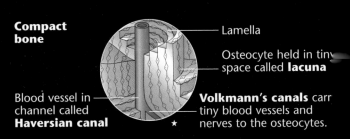

Compact bone

Blood vessel in channel called **Haversian canal** — Lamella — Osteocyte held in tiny space called **lacuna** — **Volkmann's canals** carry tiny blood vessels and nerves to the osteocytes.

Internet links

Go to **www.usborne-quicklinks.com** for links to the following websites:

Website 1 Detailed bone information, presented in a fun way.

Website 2 Take an online challenge about joints and bones.

Website 3 Watch a short movie to discover why bones are so strong.

Website 4 Find out more about joints.

Website 5 Find out about the foot bones and how they affect sports performance.

MUSCLES

Muscles are areas of stretchy tissue found all over your body. They are responsible for movement. Muscles that you can control, such as those which lift your arm, are called **voluntary muscles**. Muscles which work automatically, such as those which make your heart beat, are **involuntary muscles**. There are three types of muscles: skeletal, cardiac and visceral muscles.

SKELETAL MUSCLES

Your body contains about 640 **skeletal muscles**. These are voluntary muscles which are attached to the skeleton, usually by tough bands of tissue called **tendons**. Some skeletal muscles are attached to skin. The muscles in your face, for example, are skeletal muscles. They allow you to make different expressions.

When a muscle contracts, it shortens and tightens, pulling the bone (or skin) with it. Muscles cannot push, though, so you need another muscle to pull the part back to its original position. The muscle which contracts is called the **agonist**, and the one which relaxes is called the **antagonist**. Pairs of muscles that work in this way are known as **antagonistic pairs**.

This picture shows the main skeletal muscles.

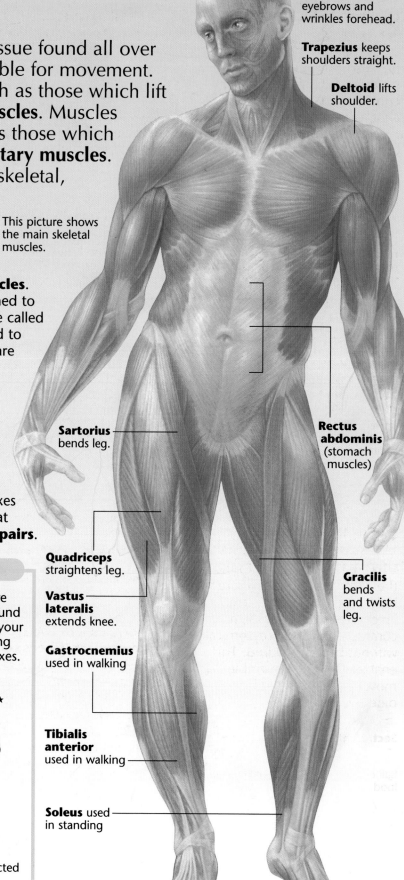

Frontalis raises eyebrows and wrinkles forehead.

Trapezius keeps shoulders straight.

Deltoid lifts shoulder.

Rectus abdominis (stomach muscles)

Sartorius bends leg.

Quadriceps straightens leg.

Vastus lateralis extends knee.

Gracilis bends and twists leg.

Gastrocnemius used in walking

Tibialis anterior used in walking

Soleus used in standing

See for yourself

The **biceps** and **triceps** muscles in your arm are an antagonistic pair. Place your hand gently around your upper arm while you bend and straighten your arm. You can feel your biceps and triceps working as a pair – one muscle tightens as the other relaxes.

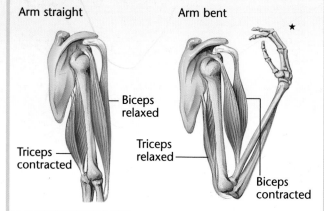

Arm straight

Arm bent

Biceps relaxed

Triceps relaxed

Triceps contracted

Biceps contracted

CARDIAC MUSCLES

Cardiac muscles make up most of your heart. They are involuntary muscles, which never tire. They form two separate sets. The upper set contracts, filling the lower chambers of your heart with blood. The lower set then contracts and squeezes the blood out into your arteries*.

Cardiac muscles at work

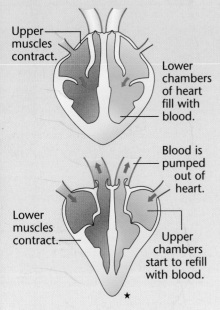

Upper muscles contract.

Lower chambers of heart fill with blood.

Blood is pumped out of heart.

Lower muscles contract.

Upper chambers start to refill with blood.

VISCERAL MUSCLES

Visceral muscles are found in the walls of many of the organs inside your body. They are involuntary muscles which contract slowly and rhythmically without becoming tired. This enables them, for example, to move food through your digestive system*.

Section of intestine

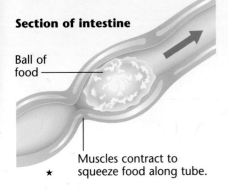

Ball of food

Muscles contract to squeeze food along tube.

* Arteries, 15; Digestive system, 18.

MUSCLE TISSUE

Skeletal muscles are made of **striated** or **striped muscle tissue**. This is made up of long, rod-shaped cells called **muscle fibers** or **myofibers**.

Myofibers are grouped into bundles called **fascicles**. Each fiber is made up of cords called **myofibrils**. These contain interlocking thick and thin threads called **myofilaments**. Thick myofilaments are made of a type of protein called **myosin**, and thin filaments are made of **actin**, another protein.

When a muscle contracts, its filaments slide past each other. This makes the muscle shorter and fatter.

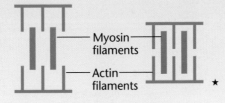

Relaxed skeletal muscle

Contracted skeletal muscle

Myosin filaments

Actin filaments

Filaments slide past each other.

Striated muscle fibers can be divided into two types. **Slow-twitch fibers** contract slowly and use relatively little energy. They can act for a long time without becoming tired. **Fast-twitch fibers** contract quickly, using more energy. They act in short, powerful bursts, but tire quickly.

The neck muscles which support your head contain many slow-twitch fibers.

Arm muscles used for throwing contain many fast-twitch fibers.

Structure of a skeletal muscle

Muscle fiber enclosed by membrane

Muscle protected by tough layer called **epimysium**

Myofibril

Fascicle protected by layer called **perimysium**

Myosin myofilament

Actin myofilament

Cardiac muscles are made of a type of striated muscle tissue called **cardiac muscle tissue**. This is made of interlocking Y-shaped fibers.

Cardiac muscle fiber

Visceral muscles are made of spindle-shaped fibers which join together to form **smooth muscle tissue**.

Smooth muscle fiber

THE CIRCULATORY SYSTEM

Your **circulatory system** transports substances, such as food and oxygen, around your body, and collects some waste substances. It has three main parts: **blood**, a liquid which carries the substances to and from the cells; tubes called **blood vessels**, through which the blood travels; and the **heart**, which pumps blood to all parts of your body.

HEART

Position of heart

Your heart is a muscular organ which, unlike other muscles, never gets tired. It is divided into four parts called **chambers**. The two upper chambers are called **atria**. They are joined to two lower chambers, called **ventricles**.

One-way valves between the chambers keep your blood flowing in the right direction. The valves have flaps called **cusps**. As blood flows through the valves, it forces open the cusps. They then snap shut, to stop blood from flowing back. As the valves shut, they make the thumping "heart beat" sound.

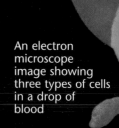

An electron microscope image showing three types of cells in a drop of blood

How blood circulates around the heart

The **aorta** is an artery which carries blood out of the heart to be taken to the rest of the body.

The **superior vena cava** is a vein which carries blood to the heart from the upper body.

Pulmonary veins carry blood from the lungs to the heart.

Arterial valves

Right atrium

Right ventricle

The **inferior vena cava** is a vein that carries blood to the heart from the lower body.

Valve open

Valve shut

Cusp

Pulmonary arteries (left and right) carry blood from the heart to the lungs.

Pulmonary veins

Left atrium

Atrio-ventricular valves

Left ventricle

Aorta

CIRCULATION

Blood passes through the heart twice during one complete circulation of your body. First, it is pumped from the right side of the heart to the lungs, where it picks up fresh oxygen you have breathed in. It then returns to the left side of the heart, from where it is pumped to the rest of the body to deliver the oxygen. Blood needing oxygen returns to the heart to begin the cycle again.

How blood travels around the body

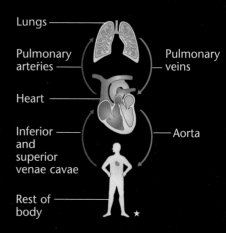

Lungs

Pulmonary arteries

Pulmonary veins

Heart

Inferior and superior venae cavae

Aorta

Rest of body

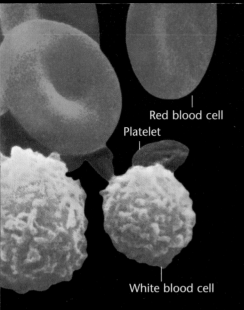

Red blood cell

Platelet

White blood cell

BLOOD VESSELS

Blood flows away from your heart along strong blood vessels called **arteries**. These branch off into ever smaller vessels, ending up with tiny tubes called **capillaries**. Their walls are only one cell thick, so oxygen and other substances needed by body cells can pass through easily into the **tissue fluid** around the cells.

Types of blood vessel

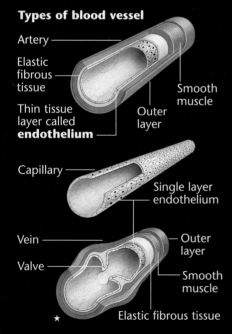

Artery

Elastic fibrous tissue

Thin tissue layer called **endothelium**

Smooth muscle

Outer layer

Capillary

Single layer endothelium

Vein

Valve

Outer layer

Smooth muscle

Elastic fibrous tissue

The tissue fluid takes substances between cells and the blood. Carbon dioxide and some waste pass into the capillaries, which eventually join up again into blood vessels called **veins**. These carry blood back to the heart.

BLOOD

Blood is made up of red and white blood cells and platelets, floating in a pale yellow liquid called **plasma**. The average adult has about nine pints of blood. As well as carrying substances around your body, blood helps to fight germs, heal wounds, and control body temperature.

Composition of blood

Plasma (55%)

White blood cells and platelets (0.45%)

Red blood cells (44.55%)

Red blood cells are disc-shaped cells, which contain a purple-red chemical called **hemoglobin**. As blood passes through the lungs, oxygen combines with the hemoglobin, forming **oxyhemoglobin**, which is bright red. As the cells deliver oxygen around the body, the oxyhemoglobin turns back into hemoglobin.

Red blood cell with oxygen

Red blood cell without oxygen

The cell's disc shape helps it to squeeze along inside tiny capillaries.

Red blood cells wear out every four months and are replaced by new ones. These are made in your bone marrow* at a rate of two million per second.

White blood cells are larger than red ones. They help your body to fight disease. You can find out more about this on page 49.

Platelets are tiny fragments of cells. They help to stop the bleeding if you cut yourself.

BLOOD CLOTTING

Most minor cuts bleed for a short time, then the blood turns into a gel-like mass called a **clot**. This is made up of sticky threads of **fibrin**, which form as a result of chemical reactions started by platelets. The clot keeps more blood from leaking out, and helps to prevent germs from entering the wound.

Clot is made up of fibrin threads.

Clot will dissolve once blood vessel is repaired.

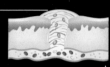

BLOOD GROUPS

Blood can be divided into four main groups – A, B, O and AB. They have different antigens* on the surface of the red cells, and different antibodies* in the plasma. In a blood transfusion, the type of blood you can be given depends on the blood group to which you belong.

Blood group	Antigen	Antibody	Can be given
A	A	Anti-B	A and O
B	B	Anti-A	B and O
AB	A and B	None	All
O	None	Anti-A Anti-B	O only

Internet links

Go to **www.usborne-quicklinks.com** for links to the following websites:

Website 1 Information about the heart, with pictures and movies.

Website 2 Detailed, interactive pictures of the cardiovascular system.

Website 3 Perform a virtual heart transplant and see a heart animation.

Website 4 Find out how blood is made and how it flows around the body, with animations.

Website 5 An interactive guide to circulation with a test-yourself activity.

TEETH

Your **teeth** help to prepare your food for digestion*. They cut and grind it up and make it easier for the rest of your body to absorb. Teeth contain living cells, nerves and blood vessels. You need to look after your teeth carefully, otherwise they will **decay** (rot away) or even fall out.

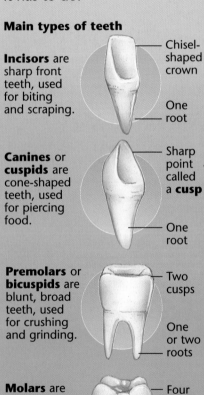

An angled mirror like this helps the dentist to see inside your mouth and check that your teeth and gums are healthy.

PARTS OF A TOOTH

A typical tooth has three main parts. The part of the tooth that can be seen is called the **crown**. Each tooth is fixed in a socket in the jawbone by one, two or three **roots**. The junction between the crown and root is called the **neck**.

Structure of a tooth

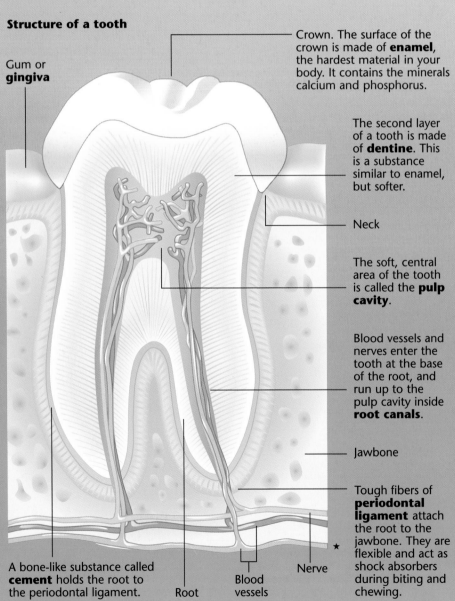

Gum or **gingiva**

Crown. The surface of the crown is made of **enamel**, the hardest material in your body. It contains the minerals calcium and phosphorus.

The second layer of a tooth is made of **dentine**. This is a substance similar to enamel, but softer.

Neck

The soft, central area of the tooth is called the **pulp cavity**.

Blood vessels and nerves enter the tooth at the base of the root, and run up to the pulp cavity inside **root canals**.

Jawbone

Tough fibers of **periodontal ligament** attach the root to the jawbone. They are flexible and act as shock absorbers during biting and chewing.

A bone-like substance called **cement** holds the root to the periodontal ligament.

Root

Blood vessels

Nerve

TYPES OF TEETH

There are four main types of teeth in an adult set. Each is shaped for the particular job it has to do.

Main types of teeth

Incisors are sharp front teeth, used for biting and scraping.

Chisel-shaped crown

One root

Canines or **cuspids** are cone-shaped teeth, used for piercing food.

Sharp point called a **cusp**

One root

Premolars or **bicuspids** are blunt, broad teeth, used for crushing and grinding.

Two cusps

One or two roots

Molars are broader than premolars, with more cusps. They crush and grind food.

Four or five cusps

Two or three roots

Wisdom teeth are the third and last molars to appear. They lie at the back of the jaw, one at each corner. They usually come through when you are 17-21.

* Digestion, 18.

This tool is called a **probe**. A dentist uses it to scrape away plaque and to check whether your teeth have holes.

TWO SETS OF TEETH

A set of teeth is called a **dentition**. During your life, you have two dentitions. The first begins to appear when you are around six months old. The teeth in this dentition are called **deciduous teeth**, **milk teeth** or **baby teeth**. There are 20 deciduous teeth in all.

Deciduous dentition (milk teeth)

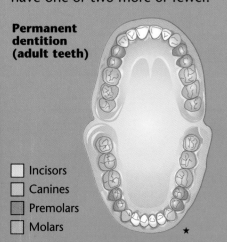

- ☐ Incisors
- ☐ Canines
- ☐ Premolars

There are no deciduous molars.

Between the ages of 6 and 12, the deciduous teeth fall out and are replaced by **permanent teeth**. There are normally 32 of these, although some people have one or two more or fewer.

Permanent dentition (adult teeth)

- ☐ Incisors
- ☐ Canines
- ☐ Premolars
- ☐ Molars

TOOTH DECAY

Everyone has tiny organisms called bacteria living in their mouth. These multiply very quickly if they have a supply of sweet foods. They form a sticky substance called **plaque**, which covers your teeth in a thin, white film.

As they feed on food lodged between your teeth, bacteria produce acid which dissolves the tooth. This can make your tooth ache, and eventually destroy it. The stages of tooth decay are described below.

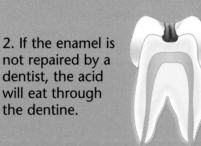

1. Bacteria feed on sweet food stuck to a tooth. The acid they produce dissolves the enamel.

2. If the enamel is not repaired by a dentist, the acid will eat through the dentine.

3. If the decay reaches the pulp cavity and its nerve endings, your tooth will start to hurt.

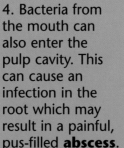

4. Bacteria from the mouth can also enter the pulp cavity. This can cause an infection in the root which may result in a painful, pus-filled **abscess**.

See for yourself

You can use disclosing tablets from a pharmacist or your dentist to see the plaque on your teeth.

Brush your teeth well, then use a disclosing tablet, following the instructions on the packet. The stained areas will show any food or plaque that you have not brushed away.

HEALTHY TEETH

Bacteria in your mouth also cause **gum disease**, or **gingivitis**. This makes the gums bleed and, if left untreated, affects the periodontal ligament and jawbone, making the teeth wobble or even fall out.

Brushing your teeth twice a day is the best way to keep your teeth and gums healthy. Many toothpastes contain a mineral called **fluoride**. This strengthens teeth by making the enamel less soluble in acid and replacing minerals in enamel that has been attacked by acid. It also reduces the bacteria's ability to make acid.

Internet links

Go to **www.usborne-quicklinks.com** for links to the following websites:

Website 1 Interactive information about teeth.

Website 2 Watch a short animated movie about teeth and try a quiz.

Website 3 Friendly information about teeth and the causes of cavities and how to avoid them.

Website 4 In-depth information on tooth decay, gum disease and tips on caring for your teeth.

Website 5 Animations about primary and permanent teeth and tooth care.

DIGESTION

As food passes through your body, it is broken down into pieces small enough to be dissolved in your blood. This process, called **digestion**, takes place in the **digestive tract** or **alimentary canal** – a tube that runs from your mouth to a hole in your bottom called the **anus**. Food is broken down physically by chewing and churning, and chemically by the action of **digestive juices**, made by organs called **glands**.

STAGES OF DIGESTION

1. Food is chewed in the mouth and mixed with a digestive juice called **saliva**, which is made in your **salivary glands**. Saliva moistens the food so it slides down your throat easily. It also starts to break down starch* in the food into a sugar called **maltose**.

2. Your throat muscles guide the food through the **pharynx** into a passage called the **esophagus** or **gullet**. As you swallow, a flap called the **epiglottis** blocks off the top of your **windpipe** or **trachea**, so the food does not go down the wrong way.

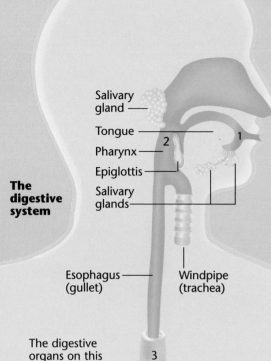

The digestive system

- Salivary gland
- Tongue
- 2
- 1
- Pharynx
- Epiglottis
- Salivary glands
- Esophagus (gullet)
- Windpipe (trachea)
- 3

The digestive organs on this diagram are shown raised and spread out so you can see them clearly.

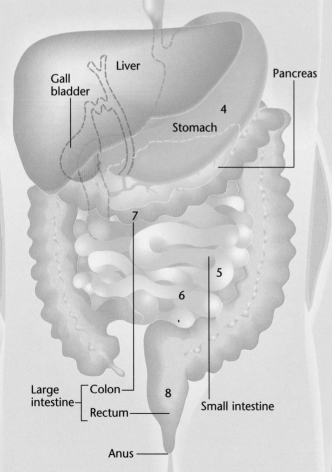

- Liver
- Gall bladder
- Pancreas
- 4
- Stomach
- 7
- 5
- 6
- Large intestine
- Colon
- 8
- Small intestine
- Rectum
- Anus

★

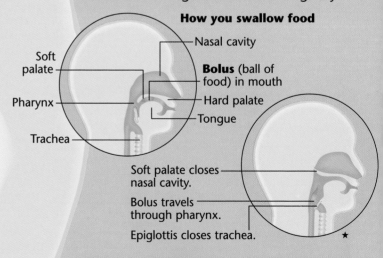

How you swallow food

- Nasal cavity
- Soft palate
- **Bolus** (ball of food) in mouth
- Pharynx
- Hard palate
- Tongue
- Trachea

- Soft palate closes nasal cavity.
- Bolus travels through pharynx.
- Epiglottis closes trachea.
- ★

See for yourself

Put a piece of crusty bread in your mouth and notice the taste as you start to chew it. After a minute of chewing, you will find that the bread starts to taste sweeter. This happens as your saliva begins to turn the starch into sugar.

* Starch, 20.

3. Food travels down the gullet to the stomach. Muscles in the wall of the gullet contract to push the food along. This action, called **peristalsis**, takes place all along your digestive tract.

4. In the **stomach**, food is churned up with **gastric juices**. These start to digest protein*, and they also contain hydrochloric acid which kills germs in the food. Your stomach lining has folds called **rugae**, which flatten as it fills.

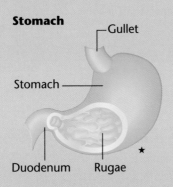

Stomach

- Gullet
- Stomach
- Duodenum
- Rugae
- ★

5. The food moves into a tube called the **small intestine**. This has three sections: the **duodenum**, the **jejunum** and the **ileum**. In the duodenum, digestive juices made by the liver and pancreas (see right) break down fats*, protein and starch*.

Small and large intestine

- Duodenum
- Colon
- Jejunum
- Ileum
- Rectum
- Anus
- ★

6. The small intestine, especially the ileum, is lined with tiny, finger-like projections called **villi** which increase its surface area. Each villus contains minute blood vessels, which absorb the digested food, and carry it to the liver for further processing before it is carried around the body.

Cross section of small intestine

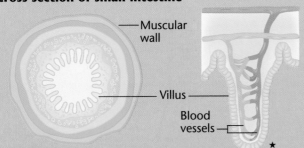

- Muscular wall
- Villus
- Blood vessels
- ★

7. Water and any food, such as dietary fiber*, that cannot be digested move into the first part of the **large intestine**, called the **colon**. There water is absorbed into your bloodstream.

8. The semi-solid waste matter, called **feces**, then passes into the second part of the large intestine, called the **rectum**. It is pushed out through the anus when you go to the bathroom.

Dietary fiber, 21; Fats, Proteins, Starch, 20.

DIGESTIVE GLANDS

Your digestive glands make fluids needed for digestion. Many digestive juices contain chemicals called **digestive enzymes**, which help to break down your food. Some digestive glands are tiny, and set into the walls of digestive organs. For example, the wall of your stomach contains **gastric glands**. Other glands, such as your salivary glands, are separate organs.

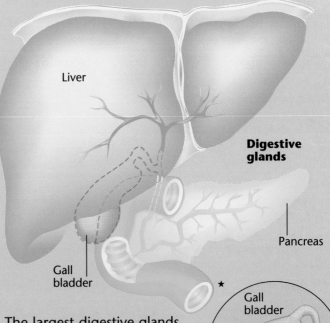

- Liver
- Digestive glands
- Pancreas
- Gall bladder
- ★

Gall bladder

- Rugae flatten as sac fills with bile.
- ★

The largest digestive glands are the liver and pancreas. Your liver makes a green liquid called **bile**. This acts like a detergent, breaking up fats* into tiny drops so that enzymes can work on them. Bile is stored in a sac called the **gall bladder**. Your pancreas makes **pancreatic juice**. This contains enzymes which break down fats, protein* and starch*. Your liver and pancreas also have other important jobs to do, for example controlling the amount of glucose in your blood. You can read more about this on page 27.

You can read more about this on page 27.

Internet links

Go to **www.usborne-quicklinks.com** for links to the following websites:

Website 1 Watch an apple travel through the digestive system.

Website 2 Take a guided tour of the digestive system and test your knowledge with an online puzzle.

Website 3 Try an online challenge and see if you can complete a 3-D jigsaw puzzle of the internal organs.

Website 4 A yucky look at the digestive system.

FOOD AND DIET

The food and drink you take in are known as your **diet**. A healthy diet consists of a variety of foods, because different foods contain different things that your body needs. Carbohydrates, proteins and fats are vital for energy or growth. They are called **nutrients**. Vitamins, minerals and water are **accessory foods**. They help your body to work properly.

CARBOHYDRATES

Carbohydrates are energy-giving foods. There are two types: sugars and starch. **Sugars** are sweet and dissolve in water. Foods such as fruit and chocolate contain sugars. **Starches** are not sweet and do not dissolve in water. Bread, pasta, potatoes and rice are rich sources of starch.

Chocolate contains sugar, a carbohydrate.

During digestion, carbohydrates are broken down into simple sugars such as **glucose**. Your body uses these as fuel to produce energy. Some glucose is converted into **glycogen** and stored in the liver. Any remaining glucose is turned into fat, and stored under the skin.

Pasta is a healthy source of starch.

PROTEINS

Proteins are used for the growth and repair of body tissue, as well as other vital jobs. They are found in lean meat, fish, eggs, nuts, milk and beans.

Proteins are made up of simpler chemical units called **amino acids**. The type of protein depends on the order in which its amino acids are arranged. During digestion, proteins are broken down into individual amino acids. These are then rearranged to make different proteins needed by your body.

Examples of proteins in the body

Hemoglobin in these blood cells carries oxygen around the body.

Keratin is the protein from which hair and nails are made.

Actin and myosin enable muscles to contract.

FATS

Fats are needed by your body for energy and warmth. Unused fats are stored in various areas of your body, such as under the skin. There are two types of fats: saturated and unsaturated.

Saturated fats are found mostly in animal products, such as butter, lard and fatty meat. These foods also contain **cholesterol**, a fat-like substance. **Unsaturated fats** are found in non-animal products, including vegetable oils and nuts.

Junk food is often high in fat. Eating too much saturated fat and cholesterol may be linked to heart disease.

See for yourself

Look at the labels on the packaging of some of the foods that you eat. They tell you how much carbohydrate, protein and fat the foods contain. Some labels also include information about the vitamins and minerals in a particular food.

VITAMINS

Vitamins are substances your body needs to remain healthy. They are found in a wide variety of foods. A balanced, healthy diet will give your body all the vitamins it needs.

Vitamins are **organic** chemicals, which means that they contain carbon. Your body needs tiny amounts of about 15 different vitamins for essential chemical processes to take place.

Vegetables and fruit are good sources of dietary fiber, vitamins and minerals.

Sources and uses of vitamins

Vitamin	Good sources	Necessary for
A (retinol)	Milk, butter, eggs, fish oils, fresh green vegetables	Eyes (especially seeing in very dim light), skin
B (a group of several vitamins)	Wholewheat bread and rice, yeast, liver, soy beans	Energy production in all your cells, nerves, skin
C (ascorbic acid)	Oranges, lemons, blackcurrants, tomatoes, fresh green vegetables	Blood vessels, gums, healing wounds, preventing colds
D (calciferol)	Fish oils, milk, eggs, butter (and sunlight)	Bones, teeth
E (tocopherol)	Vegetable oils, wholewheat bread, rice, eggs, butter, fresh green vegetables	Not yet fully understood
K (phylloquinone)	Fresh green vegetables, liver	Clotting blood

MINERALS

Minerals are another group of substances needed by your body. They are **inorganic**, which means that they do not contain carbon. You need small amounts of about 20 different minerals in all. **Trace minerals**, such as iron, are minerals that are needed in extremely small quantities. For more about vitamins and minerals, see page 55.

Sources and uses of some minerals and trace minerals

Mineral	Good sources	Necessary for
Calcium and **phosphorus**	Milk, cheese, butter, water in some areas	Strong bones and teeth
Sodium	Salt, milk and spinach	Blood, digestion, nerves
Fluorine (trace mineral)	Milk, toothpaste, drinking water in some areas	Healthy teeth and bones
Iodine (trace mineral)	Seafood, table salt, drinking water in some areas	Hormone* thyroxin
Iron (trace mineral)	Liver, apricots and green vegetables	Hemoglobin in red blood cells*

DIETARY FIBER

Dietary fiber, also known as **roughage**, is a type of carbohydrate found in bran, wholemeal bread, fruit and vegetables. Fiber cannot be digested by humans. It is also bulky, which helps the muscles of your intestines to move food efficiently through your digestive system.

WATER

Water is vital for life. Without it, you would only survive for a few days. You need to take in water to replace what you lose, for example in urine* and sweat. There is water in what you drink, and also in some solid foods, such as lettuce, which is 90% water.

About 65% of your body is water. In very young children, up to 75% of their body weight is made up of water.

*

Internet links

Go to **www.usborne-quicklinks.com** for links to the following websites:

Website 1 Check your knowledge of nutrients and play nutrition games.

Website 2 Find out why it's important to eat right, with food guide pyramid facts, recipes, and online nutrition games.

Website 3 Lots of information on healthy eating.

Website 4 Play an online game and learn about parts of the body and foods that keep you healthy.

Website 5 Explore a clickable human body and find out more about the vitamins and minerals your body needs.

Website 6 Why fruit and vegetables are good for you.

* Hormones, 26, 27; Red blood cells, 15; Urine, 26.

THE RESPIRATORY SYSTEM

The **respiratory system** is made up of your **lungs** and the passages that lead to them. You breathe air into your lungs, and oxygen from the air passes into the blood, which carries it around your body. Waste carbon dioxide passes from the blood into the lungs and is breathed out.

PARTS OF THE RESPIRATORY SYSTEM

When you breathe in, air is sucked through your nose or mouth and down a tube called the **windpipe** or **trachea**. The lining of your nose and trachea make a slippery liquid called **mucus**. This warms and moistens the air, so it can travel more easily along the passages. It also traps dirt and germs in the air. Tiny hairs called **cilia** waft the mucus away from your lungs towards your nose and throat.

Cells lining the nose   Cilia

★

Your trachea divides into two tubes, each called a **primary bronchus**. One leads to each lung. There the bronchi branch off, to become **secondary** and **tertiary bronchi**, and eventually form narrow tubes called **bronchioles**.

Each bronchiole ends in a cluster of air sacs called **alveoli**. These are surrounded by capillaries*.

Oxygen passes through the thin walls of the alveoli into the network of capillaries. Carbon dioxide in the blood, produced by cells during internal respiration*, passes into the alveoli. It is removed from your body when you breathe out.

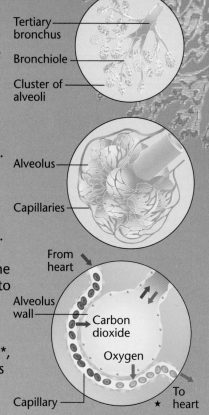

Tertiary bronchus
Bronchiole
Cluster of alveoli

Alveolus
Capillaries

From heart
Alveolus wall
Carbon dioxide
Oxygen
Capillary
To heart ★

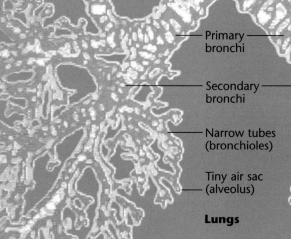

Primary bronchi

Secondary bronchi

Narrow tubes (bronchioles)

Tiny air sac (alveolus)

Lungs

Each lung contains many tubes, the smallest of which end in tiny air sacs (alveoli).

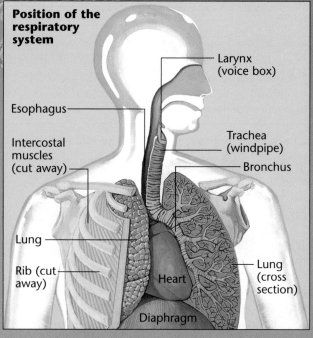

Position of the respiratory system

Larynx (voice box)

Esophagus

Intercostal muscles (cut away)

Trachea (windpipe)

Bronchus

Lung

Rib (cut away)

Heart

Lung (cross section)

Diaphragm

* Capillaries, 15; Internal respiration, 24.

BREATHING

Breathing, or **ventilation**, is the movement of air in and out of the lungs. It is controlled by the movements of muscles in your chest, called **intercostal muscles**, and a flat sheet of muscle called the **diaphragm**, which lies under your lungs.

Breathing in

Air with oxygen

Ribs move up and out.

Diaphragm flattens.

Breathing out

Air with carbon dioxide

Ribs move down and in.

Diaphragm rises.

★

As you breathe in, your diaphragm flattens and your intercostal muscles contract, pulling your ribs up and out. The space inside your chest increases and makes the pressure of the air in your lungs lower than that outside your body. Air rushes in to fill the space. This is called **inhalation**.

As you breathe out, your diaphragm relaxes upward, and your intercostal muscles relax, so your ribs move down and in. The space inside your chest gets smaller again, and air is squeezed out. This is called **exhalation**.

The normal rhythm of your breathing is sometimes interrupted. **Sneezing** clears irritating dust, pollen or germs from your nose. **Coughing** helps to clear such particles from your windpipe. **Yawning** raises the level of oxygen in your blood, and helps to rid your body of large amounts of carbon dioxide.

VOICE BOX

Your **voice box**, also called the **larynx**, is at the top of your trachea. Inside it are two bands of muscle called **vocal cords**. These open to let air past when you breathe, but when you speak or sing, muscles pull the cords together. The air passing up through the cords makes them vibrate. The vibrations can be heard as sounds.

Vocal cords as seen from above

Closed

Open

★

See for yourself

Place your fingers lightly on the front of your neck while you talk, shout and sing. You will be able to feel the vibrations in your vocal cords, and the movement of your muscles as they relax and tighten.

The louder and lower the sound is that you make, the stronger the vibrations. Your muscles will tighten when you sing higher notes and relax as you sing lower ones.

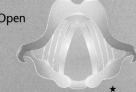

The shorter your vocal cords are and the faster they vibrate, the higher the sound you will make. Women's vocal cords are short and vibrate about 220 times in a second, so their voices are high. Men's cords are longer and vibrate about 120 times a second. This is why men have deeper voices.

Internet links

Go to **www.usborne-quicklinks.com** for links to the following websites:

Website 1 Find out more about breathing and vocal cords.

Website 2 Lots of lung information, including how to keep your lungs healthy, online games and puzzles to print out.

Website 3 Find out why we yawn then try an experiment.

Website 4 Watch a short animated movie on respiration, and try a quiz.

Website 5 Discover how mountaineers adapt to breathing air with less oxygen.

ENERGY FOR LIFE

Your body needs energy to keep alive and working. It releases energy from digested food in a series of chemical reactions. This process, called **internal respiration**, takes place in your cells, particularly in your muscles. All the processes in your body involved in producing energy, growth and waste are called **metabolism**.

A healthy diet provides this dancer with energy, and regular exercise keeps her strong and supple.

AEROBIC RESPIRATION

Internal respiration which uses oxygen is called **aerobic respiration**. Food, usually in the form of glucose*, is combined with oxygen breathed in from the air. The reaction releases energy, and its waste products are water and carbon dioxide. Chemicals called **enzymes** help to speed up the reaction.

Summary of aerobic respiration

Glucose + Oxygen

↓

Energy + Carbon dioxide + Water

Some of the energy is set free as heat in a process called **thermogenesis**. The rest is stored as a chemical called **ATP** (**adenosine triphosphate**). When energy is needed, ATP breaks down into **ADP** (**adenosine diphosphate**), releasing its stored energy.

METABOLIC RATE

The overall rate at which your body converts food into energy is called your **metabolic rate**. It varies from person to person.

People with a slow metabolic rate convert food into energy slowly. They may gain fat easily and often appear to have little energy. People with a fast metabolic rate often appear to have plenty of energy. They convert food to energy quickly and little is stored as fat.

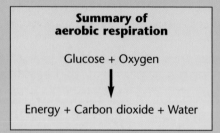

Low metabolic rate	High metabolic rate
Food	Food
↓	↓
Energy	Energy
+	
Fat	

Regular exercise, such as swimming, can help to increase your metabolic rate and keep you fitter.

ENERGY VALUE

The amount of energy that can be released from food is called its **energy value**. This is usually measured in **kilojoules** (**kJ**). Energy value is sometimes given in **kilocalories**, also known as **Calories**. A kilojoule equals 0.238 kilocalories. Most pre-packed foods have labels showing the energy value both in kilojoules and kilocalories.

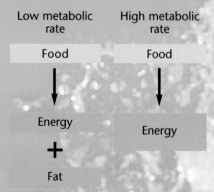

Swimming uses about 2,250 kilojoules (600 Calories) of energy per hour.

* Glucose, 20.

EFFECTS OF EXERCISE

Regular exercise is an important part of keeping healthy. It helps to keep your body fit in three ways, improving your strength, stamina and suppleness.

Strength is the amount of force a muscle or group of muscles can produce. **Stamina** helps you to exercise for longer without becoming tired. **Suppleness** describes how flexible your body is. Different types of activities help to develop these aspects of fitness. The effects of some types of exercise are shown in the table below.

BENEFITS OF EXERCISE

When you exercise, your muscles need more oxygen to release energy by aerobic respiration. You start to breathe more quickly to take in extra oxygen. This strengthens the chest muscles and increases the amount of air your lungs can hold.

Your heart beats faster to pump the oxygen-rich blood to your muscles. This strengthens the heart muscles. As blood rushes through your blood vessels, it helps to keep them clear of fatty substances that might build up and cause a heart attack.

GETTING TIRED

Often during hard exercise, such as sprinting, your body cannot take in enough oxygen for aerobic respiration. The muscles then convert glucose to energy without using oxygen in a process called **anaerobic respiration**. A substance called **lactic acid** begins to build up. Your muscles ache and your body is said to have an **oxygen debt**.

By breathing deeply after hard exercise you take in extra oxygen to "pay back" the oxygen debt.

Exercise	A	B	C
Badminton	★	★★	★
Cycling	★★★	★	★★
Dancing (energetic)	★★	★★★	★
Football	★★	★★	★★
Gymnastics	★	★★★	★★
Hill walking	★★	★	★
Horse riding	O	O	★
Jogging	★★★	★	★

Exercise	A	B	C
Judo	★	★★★	★
Roller blading	★★	O	★
Skipping (vigorous)	★★★	O	★
Swimming	★★★	★★★	★★★
Tennis	★	★★	★
Walking	★	O	O
Weightlifting	O	O	★★★
Yoga	O	★★★	O

Key   **A** = Stamina   **B** = Suppleness   **C** = Strength
O = no effect   ★ = beneficial effect   ★★ = very good   ★★★ = excellent

---

Summary of anaerobic respiration

Glucose $\longrightarrow$ Energy + Lactic acid

Swimming exercises all the muscles and is excellent for building strength, stamina and suppleness.

BALANCING ACT

For your body to work properly, conditions inside it, such as temperature and the levels of water and other chemicals, need to be kept constant. This is known as **homeostasis**. An important aspect of this is **excretion** – the removal of waste substances from your body. Chemicals called **hormones** also help to control the levels of substances inside your body.

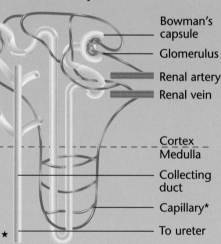

The urinary system

Right kidney

Left kidney (cross section)

Renal vein

Renal artery

Ureter

Bladder

★

Position of urinary system

★

Sphincter muscles control flow of urine out of the body.

Urethra

EXCRETORY ORGANS

Any part which removes waste from your body can be called an **excretory organ**. The main ones are your kidneys and liver, but there are others. Your lungs, for example, remove carbon dioxide and water when you breathe out, and your skin removes water and unwanted salts in the form of sweat.

This X-ray image shows the blood vessels inside a kidney.

URINARY SYSTEM

The **urinary system** controls the amount of water in your body. It is made up of two **kidneys**, a balloon-like sac called the **bladder**, and the tubes connected to them. Blood flows through **renal arteries** into the kidneys where it is filtered by about a million tiny units called **nephrons**.

Inside each nephron (see diagram, right), the artery splits into knots of capillaries* called **glomeruli**. High pressure inside the glomeruli forces glucose, water and salts to filter out of the blood into cup-shaped structures called **Bowman's capsules**.

The cleaned blood flows through the **renal vein** back to the body. The filtered liquid travels along a looping tube in the nephron, where some glucose, water and salts are reabsorbed.

Inside a nephron

Bowman's capsule

Glomerulus

Renal artery

Renal vein

Cortex
Medulla

Collecting duct

Capillary*

★

To ureter

The remaining liquid, now called **urine**, passes into a **collecting duct**, which drains into an area known as the **renal pelvis**. From there the urine flows through a tube called the **ureter** into the bladder. The urine is stored until you use the restroom, when it leaves the body through a hole called the **urethra**.

Inner part of kidney is called the **medulla**.

Outer part, where blood vessels branch out, is called the **cortex**.

The blood vessels can be seen clearly because they have been injected with a fluid which X-rays cannot pass through.

*Capillaries, 15.

HORMONES

Hormones are made in groups of cells called **endocrine glands** and are carried around the body by the blood. Your body makes over 20 types of hormones. Each type affects a different part of your body, known as its **target organ**. The main endocrine glands and some of the hormones they produce are shown in the table below.

Gland	Hormones made	Effect of hormones
Pituitary	Include growth hormone, prolactin	Control other endocrine glands, growth, mother's milk production.
Parathyroids	Parathormone	Controls calcium levels in blood and bones.
Adrenals	Adrenalin, aldosterone	Control blood glucose level, heart rate, body's salt level.
Thyroid	Thyroxin	Controls metabolism*.
Pancreas	Insulin, glucagon	Control use of glucose* by body.
Testes* (in scrotum*)	Testosterone	Controls sexual development in males.
Ovaries* (in abdomen)	Estrogen, progesterone	Controls sexual development in females.

Endocrine glands

Pituitary gland

Thyroid gland

Parathyroid glands (behind thyroid)

Adrenal gland

Pancreas

Ovary* (females only)

Testis* (males only)

*

OPPOSITE EFFECTS

Many hormones work in pairs, producing opposite effects. They are known as **antagonistic hormones**. For example, the amount of glucose* in your blood is kept at a constant level by the hormones **insulin** and **glucagon**. These are made in the pancreas by clusters of cells called **islets of Langerhans**.

If your pancreas stops making enough insulin, it causes a condition called **diabetes**. People with diabetes need to control their intake of sugar. Many also take insulin tablets or injections of insulin.

★

These three cell clusters are islets of Langerhans.

See for yourself

Some hormones act slowly, but others act very quickly. Notice the effect that the hormone adrenalin has on your body when you are excited, scared or angry. Your heart and lungs work faster to take more oxygen to your muscles. This helps to give you more power if you need to take action.

Internet links

Go to **www.usborne-quicklinks.com** for links to the following Websites:

Website 1 Lots of information about the kidneys.

Website 2 Yucky facts about urine.

Website 3 Detailed information about the endocrine and urinary systems.

Website 4 Watch a short animated movie on the endocrine system.

Website 5 Find diagrams and facts about homeostasis and hormones then test your knowledge with online tests.

How insulin and glucagon control glucose

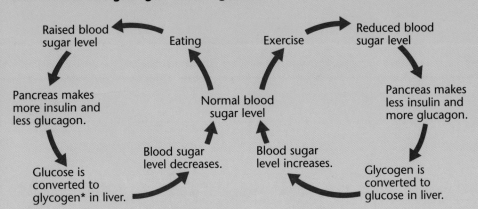

Raised blood sugar level

Eating

Exercise

Reduced blood sugar level

Pancreas makes more insulin and less glucagon.

Normal blood sugar level

Pancreas makes less insulin and more glucagon.

Glucose is converted to glycogen* in liver.

Blood sugar level decreases.

Blood sugar level increases.

Glycogen is converted to glucose in liver.

*Glucose, Glycogen, 20; Metabolism, 24; Ovaries, Scrotum, Testes, 40.

THE NERVOUS SYSTEM

The **nervous system** is made up of the brain, spinal cord and nerves. Your brain and spinal cord are known as the **central nervous system**. They receive information from all parts of your body, process it, and send instructions to other body parts. The network of nerves that carries information to and from this central area is called the **peripheral nervous system**.

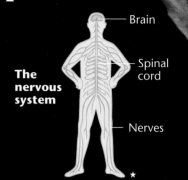

The nervous system

- Brain
- Spinal cord
- Nerves

★

NERVE CELLS

The nervous system contains millions of nerve cells, called **neurons**. There are three types: sensory, association and motor neurons.

Sensory neurons have sensitive nerve endings called **receptors**. These respond to stimuli such as light, heat or chemicals, both inside and outside the body. Sensory neurons carry information about the stimuli from the receptors to the central nervous system.

Association neurons in the brain and spinal cord pick up and interpret information from the sensory neurons. They then pass instructions to **motor neurons** which carry them to other parts of your body, such as muscles and glands, where the instructions are obeyed.

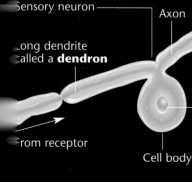

Sensory neuron —
Axon
Long dendrite called a **dendron**
Nucleus
From receptor
Cell body

Dendrites and axons may be much longer than shown in this diagram.

PARTS OF A NEURON

Each neuron has a **cell body**, which contains the nucleus, and strands called **nerve fibers**. There are two types of fibers. **Dendrites** carry information towards the cell body, and **axons** carry information away from it. The axons of one cell join another cell's dendrites, or a muscle, to pass on information.

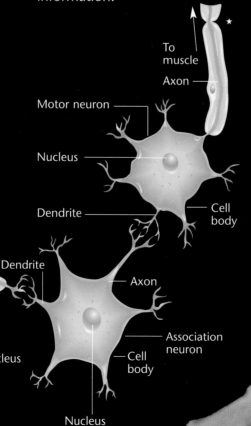

To muscle
Axon ★
Motor neuron
Nucleus
Dendrite
Cell body
Dendrite
Axon
Association neuron
Cell body
Nucleus

NERVES

Nerves are cords that contain bundles of nerve fibers. **Sensory nerves** have just the fibers of sensory neurons, and **motor nerves** have only those of motor neurons. **Mixed nerves** contain fibers from both.

Nerve
Bundle of nerve fibers
Protective sheath

The **spinal cord** is a thick bundle of nerves, which runs from the brain down a tunnel of holes in your backbone. Impulses from all parts of the body pass through the spinal cord.

In this highly-magnified picture of nerve cells in the brain, the orange areas are cell bodies.

NERVE IMPULSES

Information travels along neurons in the form of electrical signals called **nerve impulses**. When an impulse reaches the junction between one neuron and the next, a chemical called a **neurotransmitter** is released. If enough of this chemical builds up in the next neuron, an impulse is sent on.

Nerve impulse coming to end of axon

Junction called a **synapse**

Neurotransmitter builds up in end of dendrite branch.

Impulse is sent on.

TYPES OF ACTIONS

Sweating is an involuntary action.

Kicking a ball is a voluntary action.

There are two main types of actions carried out by your body. **Voluntary actions** are actions that your brain can control consciously, such as lifting a cup. Nerve impulses reach your brain and are analyzed before you decide what action to take. **Involuntary actions** are those that your brain does not consciously control. For example, processes such as digestion, breathing and circulation are involuntary. The nerves which control involuntary actions are known as the **autonomic nervous system**.

REFLEX ACTIONS

Reflex actions are involuntary actions. They are usually sudden movements, such as pulling your hand away from something hot. Most reflex actions are directed by the spinal cord. You become aware of them because other impulses are sent to the brain to tell it what is happening. The route taken by impulses during a reflex action is called a **reflex arc**.

The tangled threads on the left are nerve fibers in the brain.

A reflex arc

This shows the path taken by nerve impulses when you prick your finger.

1. Pin touches nerve endings in finger.

2. Impulses travel along sensory neuron to spinal cord.

Spinal cord (cross section)

3. Impulses travel from spinal cord along motor neuron to arm muscle.

4. Arm muscle contracts and the arm moves.

See for yourself

Try sitting on a chair with your legs loosely crossed, and tapping sharply just below your kneecap with the side of your hand. If you hit the right spot, your leg will jerk upward. This is a reflex action.

Internet links

Go to **www.usborne-quicklinks.com** for links to the following websites:

Website 1 Explore the nervous system.

Website 2 Try an online activity and learn more about the nervous system.

Website 3 Find out more about neurons and how they differ from normal cells.

Website 4 Find out why its hurts when you hit your funnybone.

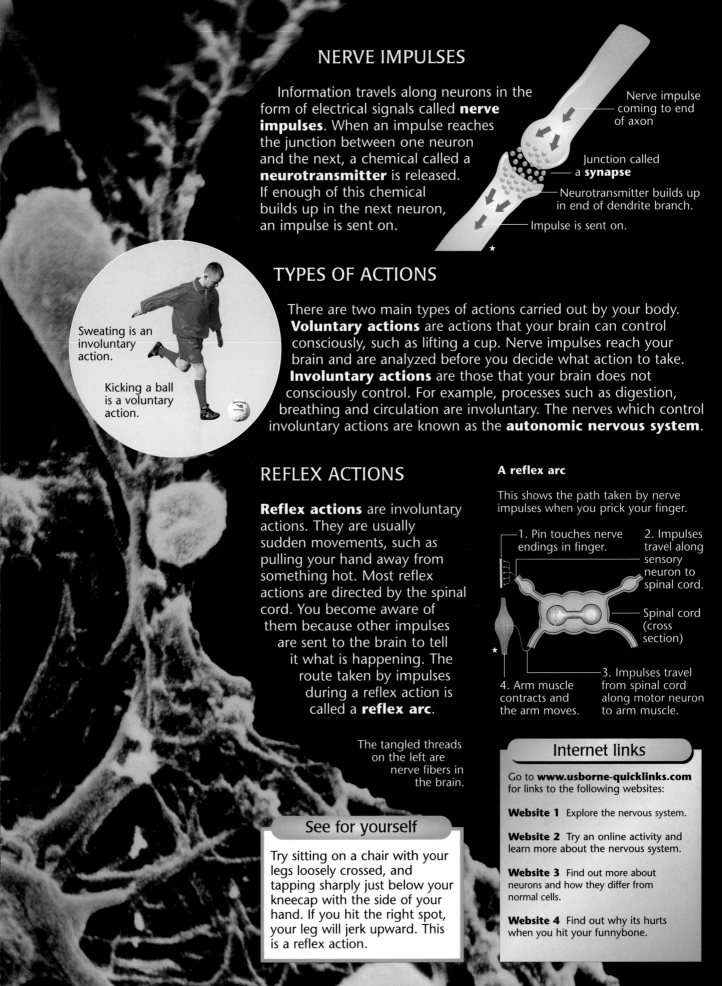

THE BRAIN

Your **brain** controls everything that happens in your body. Information, in the form of nerve impulses*, travels to and from the brain along the thick bundle of nerves in your spinal cord. The brain is the only organ that can make decisions about actions, based on past experience (stored information), present events and future plans.

INSIDE THE BRAIN

Your brain is made up of millions of neurons*. It is protected by the skull and cushioned by a thin layer of liquid called **cerebrospinal fluid**. The brain has four main parts: the cerebrum, cerebellum, diencephalon and brain stem.

The **cerebrum** is the largest part of the brain. It controls most physical activities, and many mental activities, such as thinking and learning. It also controls the **cerebellum**, which in turn coordinates muscle movement and balance.

The **diencephalon** has two parts. The **thalamus** sorts impulses as they enter the brain, and directs them to other parts of the brain for processing. The **hypothalamus** plays a vital role in homeostasis*. It controls hunger, thirst, body temperature, and the release of hormones* from the pituitary gland.

The **brain stem** controls automatic functions, such as your heartbeat and breathing. It contains three parts: the **pons**, **medulla** and **midbrain**.

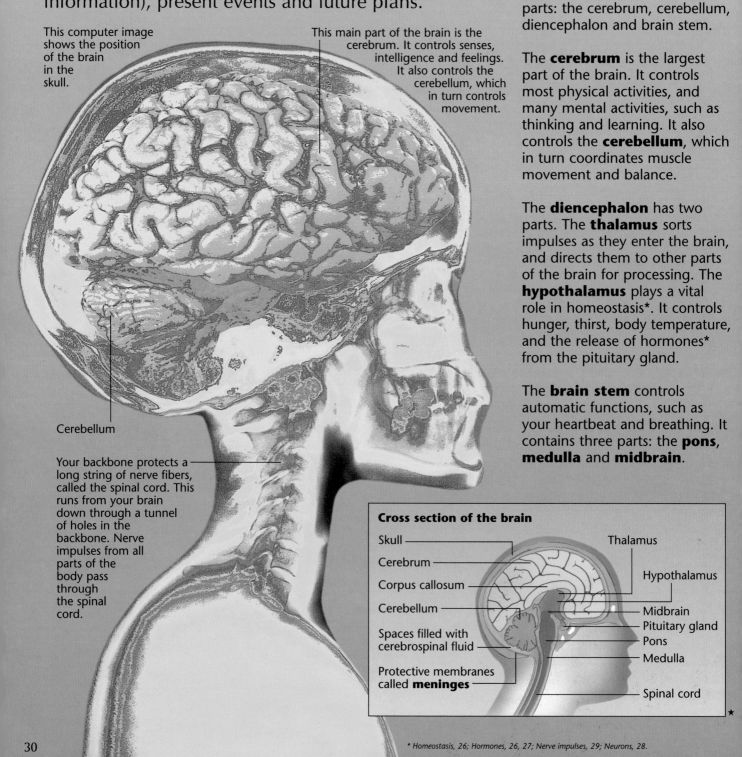

This computer image shows the position of the brain in the skull.

This main part of the brain is the cerebrum. It controls senses, intelligence and feelings. It also controls the cerebellum, which in turn controls movement.

Cerebellum

Your backbone protects a long string of nerve fibers, called the spinal cord. This runs from your brain down through a tunnel of holes in the backbone. Nerve impulses from all parts of the body pass through the spinal cord.

Cross section of the brain

Skull

Cerebrum

Corpus callosum

Cerebellum

Spaces filled with cerebrospinal fluid

Protective membranes called **meninges**

Thalamus

Hypothalamus

Midbrain

Pituitary gland

Pons

Medulla

Spinal cord

* Homeostasis, 26; Hormones, 26, 27; Nerve impulses, 29; Neurons, 28.

AREAS OF CEREBRUM

The outer layer of the cerebrum is called **cerebral cortex**. It can be divided into three types of areas. **Sensory areas** receive information from all parts of your body, such as the eyes and ears. **Association areas** analyze the information and make decisions. **Motor areas** send orders for action to muscles or glands*.

Areas in the cerebrum

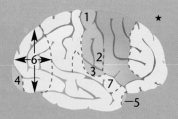

Sensory areas
1. Receives impulses from muscles, skin and inner organs.
2. Receives impulses from tongue.
3. Receives impulses from ears.
4. Receives impulses from eyes.
5. Receives impulses from nose.

Association areas include:
6. Produces sight.
7. Produces hearing.

Motor areas. Each tiny part sends out impulses to a specific muscle.

DIVIDED IN HALF

The cerebrum is made up of two halves called **cerebral hemispheres**. They are joined by the **corpus callosum**, which is a thick band of nerve fibers. Each hemisphere controls the opposite side of the body or deals with different skills.

Cerebral hemispheres

Right

Left

In right-handed people, for example, the left hemisphere controls the use of language, while the right side specializes in recognizing objects. In most left-handed people it is the other way around.

* Glands, 19, 27.

MEMORY

There are two different types of memory. **Motor-skill memory** helps you to remember how to do actions, such as walking or riding a bicycle. **Factual memory** enables you to remember specific pieces of information.

There are also two levels of memory. **Short-term memory** stores information for only a few minutes. Anything that you can remember for longer is in your **long-term memory**.

Information can be stored in your long-term memory for up to a lifetime.

See for yourself

Test your short-term memory by reading through the list of numbers below, then seeing how many you can write down in order. Most people cannot remember more than seven numbers.

3 0 9 7 1 2 8 5 4 1 6 9

BRAIN WAVES

The electrical impulses between nerve cells in your brain can be detected through your skull by sensor pads called **electrodes**. The patterns, or **brain waves**, are recorded on a chart called an **electroencephalogram (EEG)**. Doctors use EEGs to find out if a person's brain is working normally.

Main types of brain waves

Alpha waves show when you are awake, but disappear during sleep.

Beta waves show when you are thinking, or receiving impulses from your senses.

Theta waves show in EEGs of children, and adults suffering from stress or some brain disorders.

Delta waves show in EEGs of babies and sleeping adults. They can be a sign of brain disorder in an awake adult.

SLEEP

EEGs are also used to study brain activity during sleep. There are two kinds of sleep. One is known as **rapid eye movement (REM) sleep**, because your eyes move around even though they are closed. The other is called **non REM (NREM) sleep**. In REM sleep, the peaks and troughs (ups and downs) on the chart are close together, showing that the brain is very active. In NREM sleep, the peaks and troughs are further apart, so the brain is less active.

Internet links

Go to **www.usborne-quicklinks.com** for links to the following websites:

Website 1 A day in the life of a brain and what goes on inside it.

Website 2 Find lots of information, quizzes, facts and games about the brain.

Website 3-4 Probe a virtual brain to see how it controls various parts of the body and take a 3-D tour of a brain.

Website 5 Test your memory with online activities and find some tips to help you improve your memory.

Website 6 See a virtual dissection of a sheep's brain.

SKIN, NAILS AND HAIR

The **skin** is the largest organ in your body. Along with your hair and nails, it makes up the **integumentary system**. This covers your body, protecting it against damage, infection and drying out. Your skin also helps to keep your body at a constant temperature, removes some waste, makes vitamin D and helps to collect information about your surroundings.

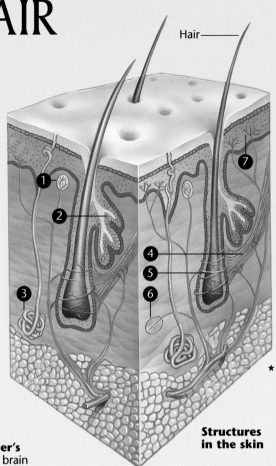

Hair

⑦

①

②

④

⑤

③

⑥

★

Structures in the skin

DIFFERENT LAYERS

Skin has two main layers: the outer **epidermis**, and the inner **dermis**. The dermis contains blood vessels, as well as structures such as receptors*. Under the dermis is a store of fat cells called the **subcutaneous layer**. This helps to keep your body warm.

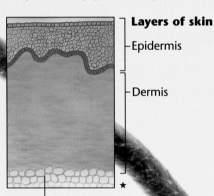

Layers of skin

Epidermis

Dermis

★

Subcutaneous layer

The epidermis has several layers. The top, **cornified layer**, is made of flat, dead skin cells, filled with a tough, waterproof protein called **keratin**. These cells are constantly being worn away and replaced by cells from a layer lower down.

Hairs growing out of skin, shown magnified over 1,000 times

* Receptors, 28.

INSIDE THE SKIN

As well as containing many blood vessels (which are not shown here), the dermis also contains other structures. These perform the skin's many jobs.

Key to structures in the skin

1. Touch receptors called **Meissner's corpuscles** send impulses to the brain when your skin touches an object.

2. **Sebaceous glands** produce an oil called **sebum**. This helps to keep your hair and skin waterproof and supple.

3. **Sweat glands** produce sweat.

4. **Hair erector muscles** make hairs stand on end, for example when your body is cold.

5. **Hair plexuses** are groups of nerve fiber endings. Each forms a network around the narrow tubes that contain hair, and sends impulses to your brain when the hair moves.

6. Pressure receptors called **Pacinian corpuscles** send impulses to your brain on receiving deep pressure.

7. **Pain receptors** send impulses to the brain if any stimulation, such as heat or pressure, becomes too much. Your brain interprets such impulses as pain.

See for yourself

Gently press a piece of clear tape onto the back of your hand, then pull it off and look at it carefully under a magnifying lens. You should be able to see tiny flakes of dead epidermal skin.

The flakes you can see here are dead skin cells from the top layer of the epidermis. They will fall off, and be replaced by cells from lower layers.

TEMPERATURE CONTROL

Your skin plays a vital part in keeping your body temperature constant, as shown below.

How skin cools down

Blood vessels widen, so more heat can be lost through the skin.

Hairs (only shown here at the surface) lie flat, so little warm air is trapped.

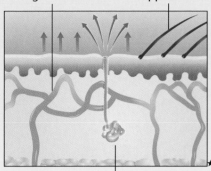

Sweat is produced. It escapes through holes called **pores**. As it dries, it uses heat from the skin, and cools you down.

How skin retains heat

Blood vessels narrow, so less heat escapes through the skin.

Erector muscles contract and make hairs stand up, trapping warm air.

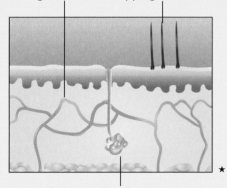

Sweat glands produce less sweat.

Your body also keeps warm by shivering. Your muscles jerk automatically, producing heat as they do so.

The outer surface of a hair is called the **cuticle**. It is made up of flat, overlapping scales of a tough substance called keratin.

NAILS

Nails help you to touch and feel things by forming firm pads to support your sensitive fingertips. Like skin, nails are mostly made from keratin. They grow from a row of dividing cells called the **nail root**.

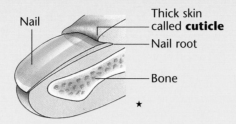

Nail

Thick skin called **cuticle**

Nail root

Bone

HAIR

Hairs grow out of deep pits, called **follicles**, in your skin. Cells at the base of each hair divide and push the hair up through the follicle. The hair you can see, called the **shaft**, is made of dead cells. This is why cutting your hair does not hurt.

How a hair grows

Hair shaft

Hair root

Follicle

Hair types

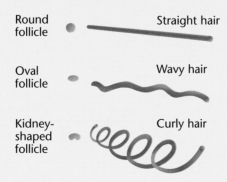

Round follicle — Straight hair

Oval follicle — Wavy hair

Kidney-shaped follicle — Curly hair

Whether your hair is curly or straight depends on the shape of each follicle.

DARK AND FAIR

Your skin contains cells called **melanocytes** which produce a brown substance called **melanin**. This absorbs some of the Sun's harmful ultraviolet rays, and so helps to protect your skin. The amount of melanin produced affects the color of the skin.

Fair-skinned people have melanin in only the lower layers of the epidermis, but people with dark skin have larger amounts of it in all layers. Melanin mixed with an orange chemical called **carotene** gives skin a yellow tint. **Freckles** are small patches of skin which contain more melanin than the surrounding area.

The color of hair is also due to melanin. Dark hair, for example, contains mostly pure melanin. Fair hair contains a type of melanin with sulfur in it, and red hair results from a type of melanin with iron in it.

Some variations in skin and hair color

Internet links

Go to **www.usborne-quicklinks.com** for links to the following websites:

Website 1 Watch short animated movies about skin, nails and hair.

Website 2 See detailed close-up images of the structure of hair.

Website 3 Find out what shampoo ingredients really do, with experiments.

Website 4 Further information about hair, nails and skin.

Website 5 A close-up look at skin.

Website 6 Learn about goosebumps.

Website 7 Try an experiment to see why metal feels cold and wood feels warm when they're the same temperature.

EYES

Your **eyes** are the organs of sight. You see things because light rays bounce off objects and enter your eyes. Light-sensitive cells at the back of your eyes send information to the brain, which interprets it as a picture, or **image**. Each eye sees objects from a different angle and your brain joins the two images to help you see in 3D. This is called **stereoscopic vision**.

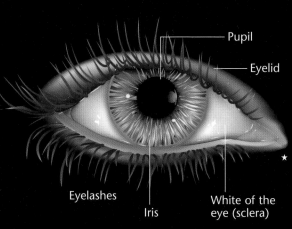

Pupil
Eyelid
Eyelashes
Iris
White of the eye (sclera)

HOW EYES WORK

Light rays enter your eye through a hole called the **pupil**. They travel through a clear layer called the **cornea** and a disc called the **lens**. These bend the light rays so that they form an image on the **retina** at the back of the eye. The lens also turns the image upside down.

The retina contains light-sensitive receptors* called **rods** and **cones**. These convert the image into nerve impulses which travel to the brain along the **optic nerve**. Your brain interprets these impulses as an image, which it also turns the right way up again.

Cross section of an eye

Gel-like fluid called **vitreous humor**

Retina

Optic nerve

Cornea helps to focus image.

Conjunctiva – a clear layer that covers the cornea

Iris controls pupil size.

Lens

Position of pupil

Fluid called **aqueous humor**

Ciliary muscles alter lens shape.

White part of eye, called **sclera**

See for yourself

There are no rods and cones on the area where the optic nerve leaves your eye. If an image falls here, you cannot see it, so this area is called the **blind spot**. Test to find your blind spot by holding this page at arm's length. Close your left eye and stare at the square with your right eye. Slowly bring the page closer to your face and notice the circle disappear.

RODS AND CONES

Each eye has about 125 million rods and 7 million cones. Rods only detect black and white, but they work well in dim light. Cones see colors but need bright light to work. At night, you see mainly in shades of gray because only your rods are working.

Close-up of area of retina

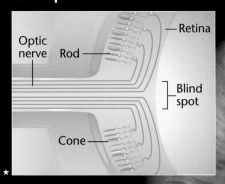

Optic nerve
Rod
Retina
Blind spot
Cone

You have three types of cones, sensitive to red, green or blue light. Each type responds by a different amount depending on the color you are looking at. For example, if you look at a purple object, the blue and red cones respond more strongly than the green ones. **Color-blind** people cannot see colors well because some cones are faulty.

* Receptors, 28.

PUPIL SIZE

The colored **iris** contains **radial** and **circular muscles** which control the size of the pupil and the amount of light entering the eye. In dim light, the radial muscles contract. This makes your pupils larger and allows in as much light as possible. In bright light, the circular muscles contract. Your pupils shrink to prevent you from being dazzled.

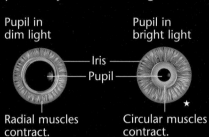

Pupil in dim light

Pupil in bright light

Iris

Pupil

Radial muscles contract.

Circular muscles contract.

Iris and pupil

The fine threads are radial muscles. They help to control the size of the pupil.

SEEING CLEARLY

As light rays from an object enter your eye, they are bent inwards by the cornea and lens. The point at which light rays meet is called the **focus**. If they focus on the retina, everything you see looks sharp and clear. The lens changes shape when looking at objects at different distances. This bends the light rays by different amounts, and keeps the image in focus.

Perfect sight

Focus falls on retina

Light rays

Some people cannot focus light properly. **Short-sighted** people cannot see distant objects clearly. They have long eyeballs, and the lens bends the rays too much, so they focus in front of the retina.

Short sight

Long eyeball

Focus in front of retina

Light rays

Long-sighted people cannot see close objects clearly. They have short eyeballs, and the lens bends the rays too little, so the image reaches the retina before it is in focus.

Long sight

Short eyeball

Focus behind retina

Light rays

Short sight can be corrected by wearing glasses or contact lenses with **concave lenses**. People with long sight need **convex lenses**.

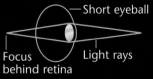

Concave lens

Convex lens

EYE PROTECTION

Eyes are very delicate. Most of the eyeball is protected by the bones of your skull. The front of the eye is protected by the thin layer of skin known as the **eyelid**.

Protecting the eyeballs

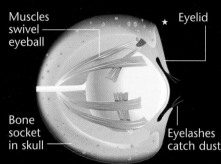

Muscles swivel eyeball

Eyelid

Bone socket in skull

Eyelashes catch dust

Eyelids keep dust and dirt out of your eyes. When you blink, your eyelids wipe **tears** over the eye, keeping it moist and clean. Tears contain chemicals which help to kill bacteria. They are made in **lachrymal glands** above each eye, and drain into your nose through two **lachrymal canals**.

Tear production in the left eye

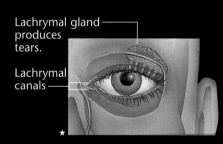

Lachrymal gland produces tears.

Lachrymal canals

Internet links

Go to **www.usborne-quicklinks.com** for links to the following websites:

Website 1 Lots of information, facts, quizzes and games.

Website 2 Useful information and an animation about eyes.

Website 3 Find out how your eyes see colors.

Website 4 Watch an online dissection of a cow's eye.

Website 5 Find out how glasses correct eyesight, then do a short experiment to see how things look through different lenses.

EARS

Your **ears** are the organs of hearing. All sounds are vibrations called **sound waves**. These enter your ears and stimulate receptors* in them to send nerve impulses* to your brain. The brain then interprets the impulses and identifies the sound. Your ears also help you to keep your balance, and give you information about the angle of your body.

As well as hearing sounds, your ears help you to keep your balance.

EARS AND HEARING

Your ear is divided into three areas: the **outer ear**, which is the part you can see, and the **middle** and **inner ear** which are the main working parts.

The ear flap, or **pinna**, funnels sound waves into a passage called the **ear canal**. The waves travel along this passage until they hit a thin layer of tissue called the **eardrum**, making it vibrate. The vibrations pass through three tiny bones (the **malleus**, **incus** and **stapes**) to the **oval window** – an oval hole covered by a thin membrane.

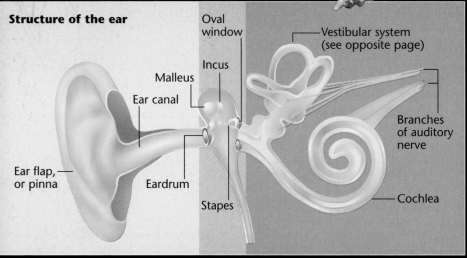

Structure of the ear

Oval window

Incus

Malleus

Ear canal

Vestibular system (see opposite page)

Branches of auditory nerve

Ear flap, or pinna

Eardrum

Stapes

Cochlea

■ Outer ear (filled with air) ■ Middle ear (filled with air) ■ Inner ear (filled with fluid)

The oval window vibrates and the vibrations pass into a spiral-shaped tube called the **cochlea**.

The cochlea contains three chambers filled with fluid. The vibrations spread through the fluid and stimulate tiny **hair cells**. These are special nerve cells attached to a membrane, called the **organ of Corti**, that runs inside the cochlea. The hair cells change the vibrations into nerve impulses, which travel along the **auditory nerve** to the brain. Your brain interprets the impulses as sounds, so you can hear.

Nerve impulses, 28; Receptors, 27.

KEEPING BALANCED

Many parts of your body help you to keep balanced. Your eyes tell you about the position of your body. So do sensitive cells known as **stretch receptors** in your muscles and tendons.

The **vestibular system** in your inner ear also has an important part to play in maintaining your balance. It has two main areas: three loops, called semicircular canals, and two sacs called the utricle and saccule.

The vestibular system

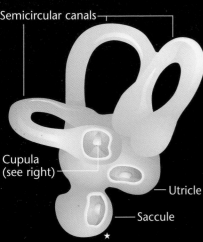

Semicircular canals

Cupula
(see right)

Utricle

Saccule

See for yourself

If you spin around very quickly, you will probably feel dizzy when you stop. This is because the liquid in your semicircular canals keeps on spinning after your body stops.

You can produce a similar effect by holding a glass of water in your hand and swirling it around.

The water in the glass will continue swirling for a little while after you stop moving the glass.

The **semicircular canals** contain fluid-filled tubes called **semicircular ducts**. At the end of each duct is a small swelling which has a gel-like projection called a **cupula** inside it. When your head turns, the fluid moves more slowly than the head, bending the cupula back. Tiny **hair cells** at the base of the cupula send your brain information about rotation of the head.

How a cupula works

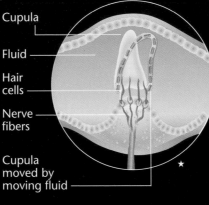

Cupula

Fluid

Hair cells

Nerve fibers

Cupula moved by moving fluid

The **utricle** and **saccule** contain a small, gel-like patch called a **macula**. It contains tiny grains, called **otoliths**, and hair cells. When your head moves, gravity causes the otoliths to slide to one side, pulling with them the gel and the hairs. These send your brain information about the forward, backward, sideways or tilted position of the head.

How a macula works

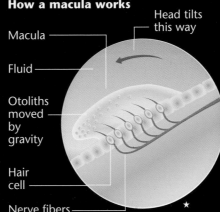

Head tilts this way

Macula

Fluid

Otoliths moved by gravity

Hair cell

Nerve fibers

TWO EARS

Having two ears gives your brain two sources of information about sounds, movement and position. By combining this information, the brain finds out more than it would from one ear alone.

For instance, having two ears helps you to tell which direction a sound is coming from. A sound coming from the left will hit your left ear slightly earlier than your right and will produce stronger vibrations. If the sound is directly in front or behind, the sound arrives at each ear at the same time and volume.

See for yourself

You can use this test to find out how your brain locates a sound. Sit on a chair, blindfolded. Ask someone to make a sound by tapping together two pencils, in different places around and above your body. Say where you think the sound is coming from.

You will probably find it hardest to pinpoint the sound when it is coming from directly behind, above or in front of you, in line with the center of your body. This is because the nerve impulses from your ears reach your brain at the same time.

Internet links

Go to **www.usborne-quicklinks.com** for links to the following websites:

Website 1 Facts and an animation about the ear.

Website 2 A simple balance experiment.

Website 3 Ear diagrams to print out.

Website 4 Interactive guide to fingerspelling, a type of sign language.

Website 5 A very detailed narrated tour of parts of the ear and how we hear.

Website 6 Yucky ear facts.

THE NOSE AND TONGUE

Your **nose** and **tongue** are the organs of smell and taste respectively. Smells and tastes are chemicals. Cells called **chemoreceptors** in your nose and tongue detect these chemicals and send information to the brain, which identifies the smell or taste. Both organs also have other important jobs to do, for example the nose is part of the respiratory system and the tongue plays a role in digestion and speech.

This man is collecting rose petals to make into perfume. The human sense of smell can detect subtle variations in perfumes.

INSIDE THE NOSE

The two holes in your nose, called **nostrils**, open into a hollow space called the **nasal cavity**. As you breathe in, air is sucked into the lower part of the nasal cavity. Here, short hairs filter out large dust particles from the air, and mucus* in the cavity's lining warms and moistens the air before it travels into the lungs.

The roof of the nasal cavity has many tiny threads called **olfactory hairs** dangling through it.

These hairs are the dendrites* of chemoreceptors called **olfactory cells**. Chemicals in the air, called **odorant molecules**, dissolve in the mucus and are absorbed by the hairs. The olfactory cells send nerve impulses* to the brain, which interprets them as a smell.

When you breathe in normally, only a small amount of air floats into your nasal cavity. When you sniff hard, you direct a stream of air toward your smell detectors. This is why things smell stronger if you sniff their scent.

DIFFERENT SMELLS

Most humans are able to recognize thousands of different scents. For many years, scientists thought that all smells were made up of the seven basic scents listed below. More recent research, however, has led to the opinion that there are many more scents – perhaps hundreds.

Seven basic scents

Smell	Example
Camphor	Mothballs
Musk	Aftershave/Perfume
Floral	Roses
Peppermint	Mint toothpaste
Ether	Dry cleaning fluid
Pungent	Vinegar
Putrid	Rotten eggs

The sense of smell is strongly linked to memory. For example, the scent of mown grass might remind you vividly of a school sports day. This link probably happens because nerve impulses from the nose are analyzed at the front of the cerebrum*. This part of the brain also deals with memory and feelings.

Inside the nose

3. Nerve impulses are carried to the brain.

2. Axons* of olfactory cells pass through the bone roof of the nasal cavity.

1. Olfactory hairs absorb dissolved odorant molecules.

Nasal cavity

* Axons, 28; Cerebrum, 30; Dendrites, 28; Mucus, 22; Nerve impulses, 29.

TONGUE AND TASTE

The main purpose of your sense of taste is to tell you whether or not something is safe to eat. For example, rotten food and most poisonous plants taste revolting, so your immediate reaction is to spit them out.

The surface of your tongue is covered with tiny bumps called **papillae**. Many of these are lined with **taste buds**, which contain chemoreceptors called **gustatory receptor cells**. These are sensitive to chemicals from your food which dissolve in your saliva. The cells send nerve impulses to the brain, which interprets them as a taste.

Taste buds

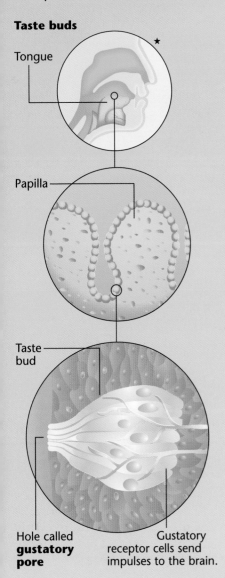

Tongue

*

Papilla

Taste bud

Hole called **gustatory pore**

Gustatory receptor cells send impulses to the brain.

Taste-sensitive areas of the tongue

These parts, called your **tonsils**, have some taste buds on them.

Bitter

Sour   Sour

Salt   Sweet   Salt

*

BASIC TASTES

Most of your taste buds lie at the sides and back of your tongue, although you have a few in other places around your throat. Buds in different areas of your tongue respond more strongly to different tastes. Scientists think that there are four main tastes: salt, sweet, sour and bitter. All flavors are made up of these basic tastes, plus smells detected by the nose.

Lemons taste sour.

Toffees taste sweet.

See for yourself

Wash your hands, and use your fingertip to place drops of cold, black coffee in different places on your tongue. Notice where your tongue is most sensitive to the coffee's bitter taste. Repeat the test using salty water, sugary water and lemon juice. Rinse your mouth with water and dry it with a piece of bread between each test.

See for yourself

You can use this test to show that the senses of taste and smell are closely linked. Grate a small amount of apple, pear and carrot into different bowls. Then shut your eyes tightly and hold your nose. Ask someone to feed you a spoonful of each food, one at a time. Try to identify the food. Repeat the experiment without holding your nose. You will probably find it easier to identify the food correctly.

TASTE AND SMELL

The senses of smell and taste are closely related. When you eat, odorant molecules from the food travel up the pharynx* into the nasal cavity, where the smell is detected in the usual way.

If you have a cold, you often lose your sense of smell and taste. This is because the lining of your nose swells, and makes thicker mucus than normal. This makes it harder for odorant molecules to reach the olfactory hairs. Your tongue can still detect the basic tastes, but you cannot identify the more subtle flavors.

Internet links

Go to **www.usborne-quicklinks.com** for links to the following websites:

Website 1 Clear information about the tongue and the nose.

Website 2 Information and experiments about taste and smell.

Website 3 Watch short animated movies about smell and taste.

Website 4 An online exhibit about our senses including taste and smell.

Website 5 Find out how we use taste and smell to experience flavor.

Website 6 Gross and interesting facts about the nose.

* Pharynx, 18.

REPRODUCTION

The process of creating new life is called **reproduction**, and the parts of the body involved in it make up the **reproductive system**. A man's body makes male sex cells, which are called **sperm**, and a woman's body produces **ova**, which are female sex cells. When a sperm joins with an ovum, a new cell is formed. This will divide many times to form a baby.

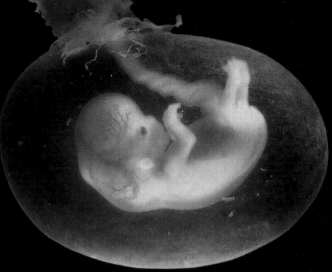

After eight weeks growing in its mother's womb, this developing baby is just over 1in long. It is floating inside a protective, fluid-filled sac called the **amniotic sac**.

MALE REPRODUCTIVE SYSTEM

Sperm are made in two organs called **testes** and are stored in a comma-shaped organ called the **epididymis**, which lies over the back of each testis. The testes sit in a sac of skin known as the **scrotum**, which hangs outside the body. The temperature inside the body would be too high for sperm to survive.

Sperm

Side view of the male reproductive system

Urethra
Penis
Foreskin
Sperm duct
Bladder
Seminal vesicle
Prostate gland
Epididymis
Testis
Scrotum
Glans
*

The **penis** is the organ through which sperm (and urine) leave the body. Its tip, called the **glans**, is very sensitive and is partly covered by a loose fold of skin called the **foreskin**. Sperm travel to the penis along two tubes called **sperm ducts**, which open into the urethra*. Several glands, including the **prostate gland** and **seminal vesicles**, make fluid in which the sperm swim. The mixture of sperm and fluids is known as **semen**.

Position of male reproductive organs

FEMALE REPRODUCTIVE SYSTEM

When a girl is born, she already has thousands of ova stored in two organs called **ovaries**. Every month, from the age of puberty*, one ovum is released from an ovary into one of the **Fallopian tubes**. This process, called **ovulation**, is described more fully on page 43.

Ovum

Front view of the female reproductive system

Fallopian tube
Ovary
Uterus
Cervix
Fallopian tube (cross section)
Vagina (leads out of body)
Ovary (cross section)
Mature ovum is released from ovary into funnel-shaped opening called **infundibulum**.
*

The Fallopian tubes lead to a hollow, pear-shaped organ called the **womb** or **uterus**. This is where a baby develops if an ovum has been fertilized (see right). At the bottom of the uterus is a muscular canal, called the **cervix**. This opens into a stretchy tube called the **vagina**, which in turn leads out of the body. The opening of the vagina lies behind that of the urethra*, and both are surrounded by two folds of skin called **labia**.

Position of female reproductive organs

MAKING BABIES

During **sexual intercourse** (also known as **sex**), the penis becomes stiff and fits inside the vagina. Muscles around the male's urethra contract, squirting a small amount of semen out of the penis into the vagina. This is called **ejaculation**.

The sperm swim up through the uterus into the Fallopian tubes. If a sperm meets an ovum, they may join together to form a **zygote** – the first cell of a new baby. This event is called **fertilization** or **conception**. If no egg is present, then the sperm die within a few days.

Fertilization

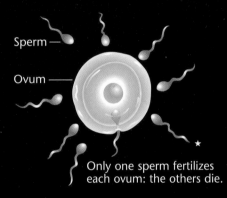

Sperm —

Ovum —

Only one sperm fertilizes each ovum: the others die.

There are several methods of preventing an ovum and a sperm from joining to make a baby. This is called **contraception**.

Internet links

Go to **www.usborne-quicklinks.com** for links to the following websites:

Website 1 View amazing month-by-month pictures of the development of an unborn baby.

Website 2 Simple animations of the menstrual cycle, fertilization and the early development of a fetus, with facts, diagrams and online tests.

Website 3 Watch a short movie about babies and reproduction.

Website 4 Fascinating sperm and egg pictures.

HOW A BABY DEVELOPS

The zygote divides to form two identical cells. These divide several times to form a ball of cells, which embeds itself in the uterus lining. The cells continue to divide, and grow into different types, such as bone or blood cells. Cells of the same type join to form **tissue**, such as muscle. Different tissues make up **organs**, such as the heart, and organs group together to form **systems**, for example the digestive system. (For more about cells, tissue and organs, see pages 8-9.)

The future baby develops over nine months. For the first two, it is called an **embryo** and for the last seven, it is a **fetus**. The mother is said to be **pregnant**.

An unborn baby gets food and oxygen from its mother's blood through an organ called the **placenta**. Waste products from the baby go back in the opposite direction. Substances pass to and from the baby through a cord called the **umbilical cord**.

At the end of pregnancy, the baby moves so that its head is near the cervix. Muscles in the uterus contract strongly, squeezing the baby out through the mother's vagina. This process is known as **labor**.

At about 40 weeks, the baby is fully developed. Approximate size: 20in

After the baby is born, its umbilical cord is clamped with a plastic clip, then cut. After about ten days, the stump falls off, leaving a **navel**.

Stages in the development of an unborn baby

The new cell formed when the ovum and sperm join together, divides in two. These two cells divide to make four, then eight, sixteen and so on until a ball of cells forms.

At six weeks, the backbone and brain are already forming. The heart starts to beat.

Approximate size: ¾ in

Umbilical cord joins embryo to placenta.

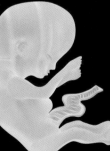

At seven weeks, tiny buds develop. They will become hands and feet.

Approximate size: almost 1in

By 12 weeks, all the organs are formed. They will develop further over the next few months.

Approximate size: 3in

GROWING AND CHANGING

During the first 20 years of life, a child gradually changes into an adult. The body grows in size and weight, and many new skills are learned. These processes are called **growth** and **development**. As you get older, your body continues to change, but more slowly. The rate at which you grow and develop depends on your genes*, as well as things such as diet and exercise.

At the age of seven, Winston Churchill's face was round and his skin was smooth.

GROWTH

Your body is made up of millions of different types of cells. To allow the body to grow, many of these divide in two to form new, identical cells. This type of cell division, called **mitosis**, also makes cells to replace many of those that wear out and die.

Parts of your body grow at different rates at different stages of life. This means that as your body grows, its proportions change. For example, a baby's head makes up one-quarter of its height, but an adult's head makes up about one-eighth of the height.

Your head also changes shape. A newborn baby has soft areas between its skull bones. Over the next few years, these are gradually replaced by bone and the head changes shape. Most parts of your body stop growing by the time you are 18, but some, such as your ears, continue to grow throughout life. Many other changes occur as you grow older, for instance your skin becomes less elastic (see photographs, right).

Child's skull ★

Adult's skull ★

At 26, his face was longer. Frown lines started to appear on his brow as his skin became less elastic.

Changes in body proportions from infancy to adulthood ★

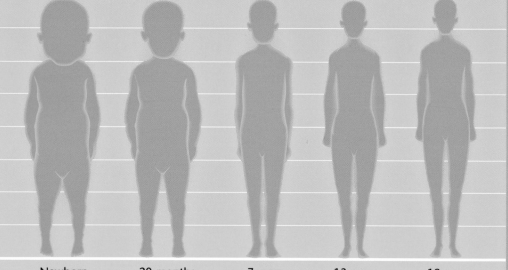

| Newborn | 20 months | 7 years | 13 years | 18 years |

By his sixties, Churchill's skin had sagged, making his face look heavier.

* Genes, 44-45.

PUBERTY

Between the ages of about 8 and 18, you change from a child into an adult. This time is known as **puberty** or **adolescence**. Changes to the body (known as **physical changes**) and to the mind and emotions (called **psychological changes**) prepare you for being an adult and a parent. These changes are triggered by hormones*.

Some of the physical changes that take place make it possible to have babies. For example, the reproductive organs* you were born with (called **primary sexual features**) become active. Other physical changes are not necessary for having babies. They result in **secondary sexual features** such as beards and other body hair.

Your feelings and emotions may change as you become more independent, explore new ways of thinking, and get used to your adult body. Changes in hormone levels in the body may also affect mood.

Physical changes at puberty

Boys	Girls
Height increases rapidly.	Height increases rapidly.
Hair grows on face; soft and downy at first, then coarser.	Fine covering of hair may grow on face.
Voice deepens.	
Hair grows under arms.	Hair grows under arms.
Shoulders and chest broaden.	Breasts start to develop.
Penis grows.	Hips widen.
Pubic hair grows around sex organs.	Pubic hair grows around sex organs.
Testes begin to make sperm.	Ovulation and periods start (see below left).

PERIODS

At birth, a girl's ovaries contain many thousands of immature ova. During and after puberty, one ovum matures about every 28 days and is released into a Fallopian tube. This is called **ovulation**. At the same time, the uterus develops a new inner layer, rich in blood vessels, ready to receive a fertilized* ovum.

If the ovum is not fertilized, this lining breaks down and leaves the body through the vagina. This is called "having a **period**" or **menstruating**. On average a girl has a period every 28 days, but it can vary. At some point between the ages of 40 and 55, the ovaries stop releasing ova and periods stop. This is the **menopause**.

AGING

After adolescence, the body becomes less efficient. This process, called **aging** or **senescence**, starts slowly but speeds up later in life. The length of time you are likely to live is your **life expectancy**. It tends to be longer if you eat healthily, exercise sensibly, avoid smoking and drug misuse, and keep your mind active.

A menstrual cycle

1. A mature ovum is released from an ovary into a Fallopian tube (ovulation). The uterus lining thickens with blood.

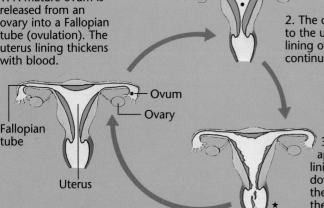

2. The ovum travels to the uterus. The lining of the uterus continues to thicken.

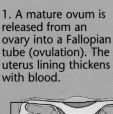

Ovum
Ovary
Fallopian tube
Uterus

3. If no fertilized ovum appears, the blood-rich lining of the uterus breaks down and passes out of the vagina, along with the unfertilized ovum.

Internet links

Go to **www.usborne-quicklinks.com** for links to the following websites:

Website 1 Good information about puberty for boys and girls.

Website 2 Explore the biology behind growing up, with online activities.

Website 3 Short animated movies about puberty, periods and aging.

Website 4 Clickable puberty diagram.

Website 5 Online exhibit about aging.

Website 6 Lots of facts and advice for girls.

* Fertilization, 41; Hormones, 26, 27; Reproductive organs, 40.

GENETICS

As soon as a sperm and an ovum join to make a new cell, that cell contains all the information needed to build a unique human being. The instructions that tell the body how to develop are **genes**, and the study of genes is **genetics**. Genes are sections of a chemical called **DNA** (**deoxyribonucleic acid**), which is packed in bundles called **chromosomes** inside a control unit called the **nucleus**. Human cells have 46 chromosomes. You inherit them from your parents.

These chromosomes are shown over 24,000 times their real size.

PAIRING UP

Your 46 chromosomes can be arranged in pairs called **homologous chromosomes**. These carry paired genes, or paired gene groups. Each gene, or gene group, on one chromosome has a partner on the paired chromosome (see opposite page).

Before cells divide for growth or repair (by mitosis*), all the chromosomes make copies of themselves, so each new cell has 46. But sex cells (ova and sperm) are made by a special type of cell division called **meiosis**. When this happens, the paired chromosomes move apart, resulting in only 23 chromosomes in each sex cell. They are ready to be paired up with new partners at fertilization*.

Inheriting chromosomes

Sperm from father — 23 → 23 → 23 → 46 → 46 46 (Cell division by mitosis) → Embryo* (46 46 46 46 / 46 46 46 46 / 46 46 46 46 / 46 46 46 46)

Ovum from mother — 23

Fertilization → Zygote* → Cell division by mitosis → Embryo*

Before chromosomes split apart to make sex cells, a certain amount of swapping of gene pairs takes place. This means that every sperm is different from every other sperm produced by the same man, and every ovum is different from every other ovum produced by the same woman. So each new child from the same parents will be different, with different genes.

GIRL OR BOY

Two chromosomes, called **sex chromosomes**, determine whether a baby develops as a male or a female. These chromosomes are known as the **X** and **Y chromosomes**. Ova and sperm have one sex chromosome each. All ova have an X chromosome. Half the sperm have an X chromosome, and half have a Y. If a sperm with an X chromosome joins with an ovum, the baby will be a girl. If a sperm with a Y chromosome joins the ovum, the baby will be a boy.

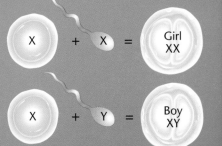

X + X = Girl XX

X + Y = Boy XY

HOW GENES WORK

You have 23 pairs of homologous chromosomes. Each gene, or gene group, on one of these chromosomes, acts together with its partner on the other paired chromosome, giving instructions to create or control one of your characteristics.

Genes for certain features, such as eye or hair color, or blood group, have different forms, called **alleles**. So a gene pair might be made up of alleles giving identical instructions, or alleles giving different instructions.

One gene may order green eyes, for example, and the other blue eyes. In such cases, either one gene will be **dominant**, overruling the other, **recessive** gene, or they will both have an effect, in which case they are called **co-dominant genes**. For example, a green eye color gene is dominant over a blue one, so if you have one of each, you have green eyes. You need two blue color genes to have blue eyes.

The diagram below shows how different pairs of blood group genes result in different blood groups, depending on which gene is dominant.

The gene for blood group A is dominant, and the gene for group O is recessive. The two people on the right are blood group A.

(A) (A)   (o) (A)

The gene for blood group B is dominant, and the gene for group O is recessive. The two people on the right belong to blood group B.

(B) (B)   (o) (B)

This person has two recessive O genes so belongs to blood group O.

(o) (o)

A and B are co-dominant so this person belongs to blood group AB.

(A) (B)

If the genes in a pair are identical, as with the AA person, the person is **homozygous** for that characteristic, in this case blood group. If they are different, the person is **heterozygous**.

Some diseases, for example a disease called **cystic fibrosis** that affects the lungs, are caused by a recessive gene. A person who has a pair of these genes will have the disease. Someone who has just one of them, paired with a normal (dominant) gene, will not be affected by the disease, but is said to be a **carrier** of that gene. The recessive gene can be passed to their children.

SEX-LINKED GENES

Some characteristics, for example color-blindness, show up more often in males than females. This is because they are caused by recessive genes found on the X chromosome which do not have partners on the Y chromosome to overrule them. Unpaired genes on the X chromosome are called **sex-linked genes**.

In genetics, genes are represented by letters. A capital letter shows that a gene is dominant, and a small one that it is recessive. The diagram below shows what might happen if a female carrier of the recessive color-blindness gene (c) has children with a man with a dominant normal sight gene (C).

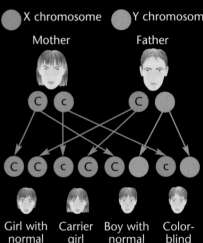

● X chromosome   ● Y chromosome

Mother   Father

Girl with   Carrier   Boy with   Color-
normal   girl   normal   blind
sight   sight   boy

Internet links

Go to **www.usborne-quicklinks.com** for links to the following websites:

Website 1 Watch an animated movie about genes.

Website 2 Advanced information, animations, videos and puzzles.

Website 3 Learn about genetics by breeding virtual mice.

Website 4 An animation about genes and disorders from the point of conception.

Website 5 A fun interactive game that shows how genes determine eye color.

GENE TECHNOLOGY

Genetic research took a great leap forward in the early 1950s, when James Watson and Francis Crick discovered the structure of DNA*. This knowledge has since helped scientists to find out much more about genes, and how living things are affected by them. New discoveries in genetics are being made all the time and these are put to a variety of uses, some of which are described here.

Watson and Crick with their DNA model

STRUCTURE OF DNA

Each molecule of DNA looks like a twisted rope ladder. This spiral shape is known as a **double helix**. The ladder's rungs are made up of four chemicals linked in pairs: adenine and thymine, guanine and cytosine. These chemicals are called **bases**, and are usually known by their initial letters, A, T, C and G.

This is part of a molecule of DNA. Its spiral shape is called a double helix.

The sides of the ladder are made up of strands of a sugar called **deoxyribose**, alternating with chemical clusters known as **phosphate groups**. One of each, together with a base, make up a unit called a **nucleotide**. A gene is a sequence of about 250 pairs of nucleotides. There are thought to be about 1,000 genes on each DNA molecule.

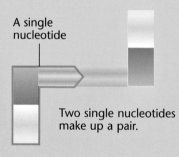

A single nucleotide

Two single nucleotides make up a pair.

The sequence of bases in a gene forms a chemical code. Each gene has a different code, and controls a different characteristic.

GENOME RESEARCH

All the DNA in an organism is its **genome**. An ordered list of all the bases in a genome is called a **map**. The genome of the yeast cell was the first to be mapped.

A significant milestone in genetic research was reached in June 2000, when scientists announced that they had made a draft of the 3.2 billion base pairs that make up the human genome. The completed map will have many uses. For example, doctors may be able to use it to find out more about the links between genes and certain diseases, and develop new ways of treating or even preventing them.

Base A is always paired with base T.

Base G is always paired with base C.

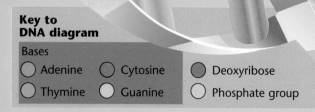

A base always joins to a deoxyribose strand.

Key to DNA diagram

Bases
- ◯ Adenine
- ◯ Thymine
- ◯ Cytosine
- ◯ Guanine

- ◯ Deoxyribose
- ◯ Phosphate group

★

* DNA, 44.

GENETIC FINGERPRINTING

Unless you have an identical twin, the exact order of bases in your DNA is slightly different from everyone else's. A process called **DNA profiling** or **genetic fingerprinting** can be used to compare samples of DNA. If the DNA samples are identical, then they have probably come from the same person or from identical twins.

DNA profiling has various uses. Police scientists, for example, can extract DNA from a single strand of hair or drop of blood left at the scene of a crime. They can then use it to identify the guilty person.

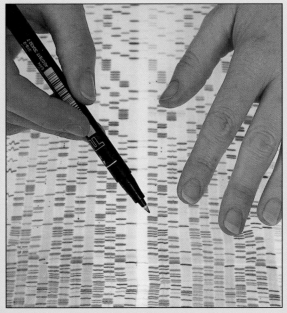

A scientist looking at DNA sequences. The bands depend on the order of bases. If the band patterns of two samples match exactly, they are likely to be either from the same person or from identical twins.

DNA samples from people who are related contain many more matching genes than samples from people who are not. Scientists can compare DNA samples to find out if and how people are related.

After the Russian Revolution in 1917, Tsar Nicholas II, his wife and three of their children were killed and buried in an unmarked grave. Bodies thought to be theirs were found in 1991. The Tsar was identified by comparing his DNA to his brother's. A DNA sample was also taken from Prince Philip, Duke of Edinburgh, a relation of the Tsar's wife. This helped to prove her identity.

GENETIC ENGINEERING

Scientists have discovered how to extract genes and use them in different ways, for instance in medicine, farming and industry. This manipulation of genes is known as **genetic engineering**. It can be used for many things, such as cloning or creating improved food crops.

The main technique used in genetic engineering is called **gene splicing**. Chemicals called **restriction enzymes** are used to cut specific genes out of DNA. Other enzymes, called **ligases**, are used to splice, or join, the genes with DNA taken from a suitable organism.

This modified DNA, known as **recombinant DNA (rDNA)**, can then be used in different ways. For example, it may be placed in a fast-breeding bacterium*. This reproduces very quickly to create lots of bacteria, each containing the rDNA with the specific gene.

A method of gene splicing

1. The gene that is needed (called the **target DNA**) is taken out of a cell's nucleus.

2. The target DNA is spliced with a **plasmid**, a special piece of DNA from a bacterium.

3. The recombinant DNA is then put into a **host bacterium** of a type which divides rapidly.

4. The host bacterium divides many times, creating many identical copies, each containing the target DNA (the desired gene).

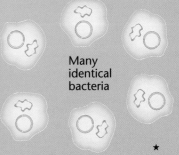

Target DNA

Plasmid

Cell

Recombinant DNA

Host bacterium

Other DNA

Many identical bacteria

*

* Bacteria, 48.

FIGHTING DISEASE

Anything that stops all or part of your body from working properly can be called a **disease**. Some diseases are caused by harmful, microscopic organisms known as **germs**. Others can be caused by diet, lack of exercise, faulty genes*, old age, or poisonous chemicals, such as nicotine from cigarettes.

White blood cells such as this help to defend your body from infection.

GERMS

The scientific name for germs is **pathogens**. Two types of pathogens – bacteria and viruses – are responsible for many diseases in humans.

Bacteria are microscopic organisms found everywhere. Harmful ones produce poisonous waste chemicals, called **toxins**, which can cause disease. Different bacteria cause different illnesses.

Viruses are strands of DNA* inside a protective coat. They cannot live on their own, but invade cells in your body and use them as factories to make more viruses. This eventually kills the cell. Diseases caused by viruses include colds, flu and AIDS.

A virus

Protective coat

Strands of DNA

★

Extension called a **pseudopodium** ("false foot") will engulf germs and trap them.

Main types of bacteria

Cocci are sphere shaped. They cause most throat infections.

Bacilli are rod shaped. They cause tuberculosis and typhoid.

Vibrios are bent rod shaped. They cause diseases such as cholera.

Spirilla are spiral shaped. They cause diseases such as ratbite fever.

★

DEFENDING THE BODY

Germs are **infectious**, which means that they pass from one living thing to another. They can be spread in a number of ways, for example in air and water and by touch. They can also be carried by animals.

Germs stick to a fly's feet and hairy body as it feeds on dung or rotten matter. These germs can spread to food a fly lands on.

Your body has many ways of defending itself from germs. Firstly your skin tries to keep germs out. But if they do get inside, your body has several methods of fighting back. The main ones are shown in the table on the right.

Your body's defences	
Skin	Forms germ-proof barrier.
Nose	Hairs and mucus trap germs and dirt from the air.
Ears	Wax inside traps germs.
Eyelids	Keep germs out of your eyes.
Tears	Wash your eyes clean.
Stomach	Hydrochloric acid kills germs in food.
Tonsils and adenoids	Kill germs in your throat.
White blood cells	Destroy germs inside your body.
Spleen	Contains white blood cells which fight infection.

This cluster of harmful germs is about to be engulfed by the white blood cell (early stage of phagocytosis – see right).

WHITE BLOOD CELLS

White blood cells move out of your blood (through capillary* walls) into tissue fluid* and lymph, and travel around fighting disease. There are two main types: monocytes and lymphocytes. **Monocytes** surround germs and digest them. This process is known as **phagocytosis** (see left and below).

Late stage of phagocytosis

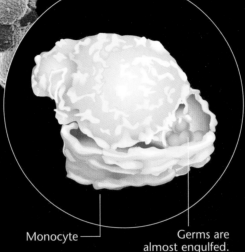

Monocyte —   Germs are almost engulfed.

Some monocytes, called **wandering macrophages**, are constantly moving around. Others, called **fixed macrophages**, become fixed in a particular organ, such as a lymph node, and fight any germs that gather there.

Lymphocytes are mostly made in lymph nodes. They destroy germs with chemicals called **antibodies**. Each type of antibody is specially produced to attack a particular chemical, or **antigen**, carried on an invading germ.

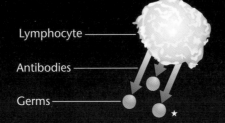

Lymphocyte —

Antibodies —

Germs —

IMMUNITY

Once you have made an antibody against a particular germ's antigen, you can make it again very quickly if the same germ enters your body. This gives you **resistance**, or **active immunity**, to the disease.

You can also be made immune to diseases such as measles by being given a **vaccine**. This is a dose of the germ that is too weak to cause the disease, but has enough antigens to make you produce antibodies. These will protect you against future attacks by the germ. This process is called **vaccination**.

Vaccination methods

In some countries, drops of vaccine against the disease polio are given on lumps of sugar.

Most vaccinations are given by injection. This stops them from being destroyed by digestive juices*.

An injection of antibodies after a disease has developed will give you **passive immunity**. The harmful germs are killed, but the immunity does not last.

Internet links

Go to **www.usborne-quicklinks.com** for links to the following websites:

Website 1 Facts about germs and the diseases they cause.

Website 2 Play a game to help a body fight back against disease.

Website 3 A short cartoon about the discovery of the polio vaccine.

Website 4 Find out about microbes, such as bacteria, and solve microbe mysteries.

Website 5 Read about the differences between viruses and bacteria.

LYMPHATIC SYSTEM

Your **lymphatic system** and your white blood cells form a disease-fighting partnership. The lymphatic system is a network of tubes and connected organs. The tubes contain **lymph**, a liquid made up of waste soaked up from tissue fluid* and white blood cells.

Lymph vessels carry lymph around your body.

Lymph drains back into your blood through two veins near your neck, recycling the white blood cells to begin again (see right). **Lymph nodes** are small organs found in clusters around the system (in your neck, armpits and groin). Many of your white blood cells are made in the lymph nodes, and many germs are trapped and destroyed there.

MEDICINE

Your body's defenses are often strong enough for you to get better without seeing a doctor. If you do need help, though, there is a whole area of science that specializes in treating disease as well as keeping the body fit and healthy. This is called **medicine**. Major advances in medicine have helped to increase the life expectancy of many people.

DIAGNOSIS

When you visit a doctor because you are ill, the doctor will ask questions and examine you to find out what is wrong. This is called making a **diagnosis**. If the doctor needs more information, there are many tests that can be done. Some of these are simple and others involve expensive and complex equipment.

Chemical analysis of samples of body fluids such as blood and urine can reveal important clues. For example, glucose in urine can be a sign of diabetes. This can be tested for by dipping a chemical-soaked stick into a urine sample. Different amounts of glucose turn the tip of the stick particular colors.

A urine test chart

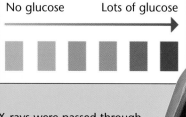

No glucose	Lots of glucose

X-rays were passed through this hand to form an image on a photographic plate.

Various methods of **medical imaging** allow doctors to see inside a patient's body without cutting it open. For example, invisible rays of energy called **X-rays** can pass through soft tissue, but not through denser substances such as bone. X-rays are particularly useful for finding out if bones are broken.

Soft areas, such as the digestive tract*, can be examined by filling them with a **radio-opaque** liquid. This keeps the X-rays from passing through, and so any blockages or changes from the usual shape can be seen clearly.

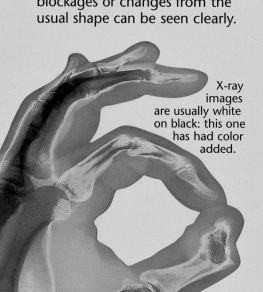

X-ray images are usually white on black: this one has had color added.

This MRI scan shows a section of the head. The walnut-shaped part is the brain and the pink blobs are eyeballs.

CT (computed tomography) scanners are special X-ray cameras that take detailed images of hard and soft tissues. The body is scanned in sections and the images are fed into a computer. Doctors look at the pictures to see if there are any unusual shadows or changes of shape that may be a sign of problems, such as abnormal growths called **tumors**.

Magnetic resonance imaging (MRI) scanners also scan sections of the body, but they use radio waves in the presence of a strong magnet. A computer builds up the images to create a 3-dimensional picture. MRI scans are used particularly to look for diseases of the nervous system and brain.

* Digestive tract, 18.

TREATMENT

Treatments range from rest, exercises or changes of diet to medicines or other, more complex, methods. In some cases, for example, a doctor may perform an **operation**, cutting the body open to mend or take out a diseased part.

MEDICINES

Chemical substances called **medicines** or **drugs** are used to treat and prevent a wide variety of illnesses. Most medicines are made in laboratories. Many are based on substances in plants which have healing properties.

Foxglove leaves contain a substance called **digitalis**. This is now made artificially and is used to treat heart disease.

Medicines called **antibiotics** are used to treat many illnesses caused by bacteria*. They either stop the bacteria from multiplying, or destroy them completely. Antibiotics have no effect on illnesses caused by viruses*, such as colds and flu.

All medicines are dangerous and should not be touched without the advice of a doctor or pharmacist. Misuse of medicines could make you very ill or even kill you.

This green furry growth is a mould called Penicillium. In 1928, Scottish scientist Alexander Fleming discovered that it could kill bacteria. He used it to develop **penicillin**, the first antibiotic drug.

SURGERY

All operations are part of the area of medicine called **surgery**. They are usually carried out in hospital by specially-trained doctors called **surgeons**. There are many different fields of surgery, each with specialized techniques.

Laser surgery uses intense beams of light, called **laser beams**, to make clean, precise cuts and to carry out delicate surgery, such as eye operations. For example, if a patient's retina* has become detached, a laser can be used to weld it back in place with a tiny heat scar. Lasers were originally developed for non-medical purposes, such as cutting and welding in industry.

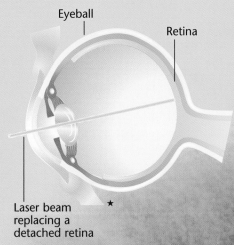

Eyeball

Retina

Laser beam replacing a detached retina ★

Lasers are often used with a tube called an **endoscope** which is pushed into a patient's body, often down the throat. Endoscopes are used to see and remove things, such as growths, inside the body.

Many endoscopes contain **fiber-optic cables**. These are made up of hair-like glass strands called **optical fibers**, through which light and laser beams can pass. Other types of cables have different jobs, such as sucking out samples for analysis.

Optical fibers in a fiber-optic cable ★

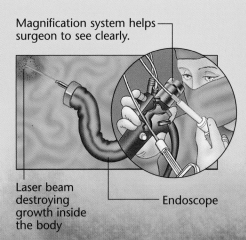

Magnification system helps surgeon to see clearly.

Laser beam destroying growth inside the body

Endoscope

Internet links

Go to **www.usborne-quicklinks.com** for links to the following websites:

Website 1 Play an activity to see how doctors treated the same disease at different times in the 20th century and read about medical discoveries.

Website 2 Explore an interactive animation on artificial replacement body parts.

Website 3 Watch detailed online tutorials about diagnostic techniques including MRI and ultrasound.

Website 4 View amazing cross-section scans of the human body.

Website 5 How the use of lasers in surgery has changed modern medicine.

* Bacteria, 48; Retina, 34; Viruses, 48.

OTHER TREATMENTS

There are many treatments, called **alternative treatments**, that have not traditionally been used by doctors. Some of these are now often used in addition to conventional treatments and are known as **complementary medicine**. Many people also use complementary treatments as part of a general healthy lifestyle. Some well-known treatments are described here.

HOMEOPATHY

Homeopathy was founded about 200 years ago by a German doctor named Samuel Hahnemann. It is based on the idea that a substance that causes certain symptoms in a healthy person can be used to cure the same symptoms in someone who is ill. It is thought that this works by helping to stimulate the body's natural defenses (see pages 48-49).

Homeopathic remedies work best in very small doses. Some are made from natural ingredients, such as herbs. Others contain minute amounts of conventional medicines.

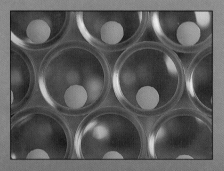

These pink homeopathic pills are plain pills that have been soaked in very dilute solutions of a remedy.

ACUPUNCTURE

Acupuncture is an ancient Chinese treatment based on the idea that all things contain an energy force called **Chi**. Chi is said to flow along invisible channels in the body called **meridians**. On these channels are hundreds of invisible points known as **pressure points**.

The meridians are shown in red. The tiny dots are pressure points.

★

An **acupuncturist** acts on these points mainly by sticking very fine needles into them. The needles do not hurt because the acupuncturist has been taught exactly where and how to stick in the needles. The acupuncturist might also act on the points with finger pressure or heat from burning herbs.

Acupuncture is used for many reasons, such as relieving pain and reducing stress. Treatment is not always in the same area as the illness. For example, the meridian that affects the lungs starts on the chest and ends at the top of the thumb. Any point on that meridian could be used to treat a lung problem depending on the diagnosis.

MANIPULATING JOINTS

Osteopathy and **chiropractic** are ways of treating physical problems by manipulating the joints, particularly the backbone. The treatments are often used for back problems, but osteopaths and chiropractors believe that other illnesses, such as headaches and rashes, can also be treated by their methods.

This pose is based on a yoga posture. Practicing yoga can help to make your body more supple.

YOGA

Some people do **yoga** to help them to relax. It combines special movements and positions, called **postures**, with techniques of breathing and concentration called **meditation**. Many people find that yoga improves their general mental and physical health as well as helping to relieve aches and pains

PREVENTATIVE MEDICINE

Preventing illness is just as important as treating it. Doctors, medical scientists and health officers spend much of their time looking for ways to control and wipe out illness. This is called **preventative medicine**.

One important way of preventing diseases is **vaccination** (see also page 49). Babies and children are usually given a series of vaccines against diseases such as polio and measles. If you go overseas, you may need a vaccine against diseases found in the country you are visiting.

Regular medical check-ups to look for early signs of disease are carried out in schools and clinics. This is called **screening**, and helps because doctors may spot and be able to treat an illness before it develops.

HEALTH ADVICE

To help keep people healthy, doctors teach them about the benefits of regular exercise and a balanced diet. They also provide information about the harmful effects to the body of smoking, drinking too much alcohol, and misusing drugs.

DRUG INFORMATION

A **drug** is any substance that affects the way the body works. Different drugs have different effects. Medicines, alcohol, and nicotine in cigarettes are all drugs. These are available legally, though their use is controlled, for example by prescriptions or age restrictions. Other drugs, such as heroin, are only available illegally.

Too much of any drug, legal or illegal, can cause long-term damage or even death. Many drugs are **habit-forming**, so people feel that they need to take them. Some are **addictive**, which means that the body gets used to them and is disturbed without them. Below is a list of some drugs and their effects.

Substance	Description	Effect on the body
Alcohol	Clear liquid found in beer, cider, wines, spirits and alcoholic "soft drinks".	Relaxation, confidence or depression. Poor coordination and judgement, so drinking and driving is very dangerous. Addictive. Long-term effects of heavy drinking include serious liver damage.
Cannabis	Often dried leaves or brown solid lump. Usually mixed with tobacco and smoked.	Relaxation. Tired, dizzy or sick feeling, with dry mouth, red eyes and faster heart rate. Combined with effects of nicotine (see below).
Cocaine	Fine, white powder. Usually sniffed.	Alert, excited or aggressive feeling. Destroys nasal passages and damages lungs. Highly addictive.
Crack	A form of cocaine. Small lumps. Smoked.	
Ecstasy	Tablets or capsules. Swallowed.	Feelings of energy and confidence or sickness and anxiety. Damages liver and kidneys. Can kill without warning.
Heroin	Gray-brown powder, often sold mixed with bleach or talc. Smoked, sniffed or injected.	Feeling of well-being then depression. Very addictive. Body needs ever higher doses or suffers awful withdrawal pains. Death by overdose is common.
Inhalants	Include lighter fuel, glue, paint or varnish. Usually sniffed.	Fumes cause feeling of well-being and dizziness. Damage lining of nose and lungs. Can suffocate. Often addictive.
LSD	White tablet or on small pieces of paper. Usually swallowed.	Puts user in strange, sometimes terrifying world, called a **trip**. Causes mental problems and brain damage.
Nicotine	In tobacco, for example in cigarettes.	Enjoyment or sick feeling. Habit-forming and addictive. Damages lungs and cilia*. Causes heart disease and chest infections. Can lead to lung cancer.

Internet links

Go to **www.usborne-quicklinks.com** for links to the following websites:

Website 1 Take a quiz to test your knowledge of various ailments, diseases and treatments.

Website 2 Information and online activities about smoking, alcohol, drugs and other health issues.

Website 3 Browse a database of health questions and answers or email your own question.

Website 4 A simple yoga sequence with clickable pictures for more information.

Website 5 More on complementary medicine including homeopathy and acupuncture.

Website 6 Travel back in time and learn about advancements in science that have helped control disease.

Website 7 Make your own vaccines online.

* Cilia, 22.

FACTS AND LISTS

DISEASES AND INFECTIONS

On this page, you can find information about various diseases and infections, including their symptoms and how they are spread. An entry in *italics* is the name of a specific bacterium or similar organism. "Droplet infection" refers to diseases that are spread through the air, for example by coughs or sneezes.

NAME	CAUSED BY	TRANSMISSION	SYMPTOMS
AIDS (Acquired Human Immune Deficiency Syndrome)	Human Immunodeficiency Virus (HIV)	Carried in blood or other body fluids; usually transmitted by sexual intercourse	Fever, tiredness, weight loss, diarrhea, viral infections
Appendicitis	Blockage in the appendix	Not transmitted	Abdominal pain, nausea
Chicken pox	*Varicella zoster* virus	Droplet infection	Blister-like rash, tiredness, headaches, sore throat
Common cold	Rhinoviruses in the nose	Droplet infection	Sneezing, blocked nose, sore throat, runny eyes
Conjunctivitis (pinkeye)	Virus, bacteria or allergy	Variable	Watery or sticky yellow discharge from the eyes
Diphtheria	*Corynebacterium diptheriae*	Droplet infection	Swelling in throat; bacteria give out toxin which can badly damage the heart
Gastro-enteritis	Bacteria, viruses and food poisoning	Droplet infection or food	Varies from nausea to severe fever, vomiting and diarrhea
German measles (rubella)	*Rubivirus*	Droplet infection	Sore throat, followed by red rash and enlarged lymph nodes
Glandular fever	Epstein-Barr virus	Saliva of infected person	Sore throat, fever, tiredness, depression
Influenza ('flu)	Influenza A, B or C virus	Droplet infection	Fever, sweating, aching muscles
Laryngitis	Adeno and rhinoviruses	Droplet infection	Sore throat, coughing, hoarseness
Legionnaire's disease	*Legionella pneumophila* bacterium	Infected or stagnant water in cisterns, cooling towers or dehumidifiers	Pneumonia and 'flu-like symptoms; fever, diarrhea, mental confusion
Malaria	*Plasmodium falciparum, P. vivax, P. ovale, P. malariae*	Bite of the anopheline mosquito	Severe fever, cold sweats, shivers
Meningitis	Various bacteria, viruses or fungi	Droplet infection	Severe headache, stiffness, nausea, sensitivity to light, rash
Mumps	*Rubulavirus*	Droplet infection	Tiredness, fever, painful jaw
Pneumonia	*Streptococcus pneumonia* bacterium	Droplet infection	Cough, fever, chest pain
Poliomyelitis (polio)	Three types of polio virus	Mostly droplet infection	Headache, fever, stiffness; may result in paralysis
Rabies	*Lyssavirus*	Saliva of infected animal	Headaches, sickness, fits, coma; usually fatal if untreated
Shingles	*Herpes zoster* virus	Dormant virus in body is activated by infection	Pain, numbness, blisters
Sleeping sickness	*Trypanosoma brucei gambiense* or *rhodesiense*	Bite from infected tsetse fly	Fever, swollen lymph glands, headache, coma; can be fatal
Tetanus	*Clostridium tetani*	Bacteria from soil infects open wound	Muscular spasms cause lockjaw, leading to breathing difficulties
Tonsilitis	Usually cold viruses	Droplet infection	Inflamed tonsils, sore throat
Tuberculosis (TB)	*Mycobacterium tuberculosis* bacterium	Inhalation of bacterium, contact with infected milk	Coughing up blood, weight loss, chest pain
Typhoid	*Salmonella typhi*	Bite from infected flea, mite, louse or tick	Fever, headaches, rash, muscular pain
Whooping cough	*Bordetella pertussis*	Droplet infection	Severe coughing, breathing difficulties

VITAMINS AND MINERALS

Here are summaries of the main vitamins, minerals and trace minerals, with information on the problems caused when they are lacking from the diet.

Vitamin/chemical name		Deficiency symptoms
A	Retinol (carotene)	Night blindness, rough skin, impaired bone growth
B$_1$	Thiamin	Beri-beri (paralysis and weakness), impaired muscle and memory function
B$_2$	Riboflavin	Blurred vision, cataracts
B$_3$	Niacin	Pellagra (dementia, skin complaints and diarrhea)
B$_6$	Pyridoxine	Facial skin problems
B$_9$	Folic acid	Anemia (blood disorder)
B$_{12}$	Cyanocobalamin	Anemia, brain disorders
C	Ascorbic acid	Scurvy (diseased gums)
D	Calciferol	Rickets and other bone defects
E	Tocopherol	Impaired fat absorption
K	Phylloquinone	Blood does not clot correctly

Mineral	Deficiency symptoms
Calcium	Bone defects
Chromium	Adult-onset diabetes
Copper	Anemia
Fluorine	Tooth decay, possibly bone defects
Iodine	Goiter (enlargement of the thyroid gland)
Iron	Anemia
Magnesium	Irregular heart beat, muscular weakness
Manganese	Unclear in humans
Molybdenum	Unknown in humans
Phosphorus	Muscular weakness, bone pain, appetite loss
Selenium	Heart problems
Sodium	Impaired acid-base balance in body fluids (very rare)
Zinc	Impaired wound healing, appetite loss

PROTEINS

Here is a list of the different types of protein found in humans, classified by their function in the body.

Enzymatic proteins
Most of the chemical reactions which take place in a living body are catalyzed (made faster or slower) by proteins called enzymes. **Amylase**, which breaks down starch in the mouth, is an example of an enzyme.

Hormonal proteins
Hormonal proteins coordinate and regulate biological processes. **Insulin**, for instance, helps to regulate the concentration of sugar in the blood.

Defensive proteins
These proteins protect against diseases. **Antibodies**, which combat bacteria and viruses, are kinds of defensive proteins.

Receptor proteins
Receptor proteins are built into the cell membrane. They recognize and bind to outside molecules such as hormones, and are involved in the cell's response to chemical stimuli.

Contractile proteins
Contractile proteins play an important part in movement. **Actin** and **myosin**, for instance, enable muscles to contract.

Structural proteins
These form an essential part of the body's supportive structure. For example, **collagen** and **elastin** provide a fibrous framework in connective tissues such as tendons and ligaments.

Transport proteins
These are involved in transporting other substances around the body. For example **hemoglobin**, the iron-containing protein in blood, transports oxygen from the lungs to other body parts.

Storage proteins
Some substances are stored in the body by combining with proteins. For instance, iron is stored in the liver with the protein **ferritin**.

Nutrient proteins
These are specialized storage proteins for amino acids. **Casein**, a protein in milk, is a major amino acid source for baby mammals.

A-Z OF SCIENTIFIC TERMS

Aerobic respiration Internal respiration in which oxygen is combined with glucose to produce energy. The products are water and carbon dioxide.

alimentary canal See *digestive tract*.

allele Any of the different forms in which a gene can occur. For example, an eye color gene might be a blue or green allele.

alveoli Tiny sacs at the end of bronchioles, where gases are exchanged between the lungs and the blood.

amino acids Fatty acids which join together in chains to make up proteins.

ampulla See *semicircular ducts*.

anaerobic respiration Internal respiration which may occur during hard exercise, when aerobic respiration is not fast enough to produce all the energy needed. Glucose in the muscles is broken down directly for energy (without using oxygen), and **lactic acid** is built up as a result, creating aching muscles and an **oxygen debt**. This is repaid later by breathing deeply to take in extra oxygen.

antagonistic hormones A pair of hormones producing opposite effects, which work together to balance chemical levels.

antagonistic pair Two muscles which work together to move a body part. The part is moved by the **agonist**, which contracts while the **antagonist** relaxes. When returning the part to its original position, the muscle that was the antagonist becomes the agonist, and vice versa.

antibodies See *lymphocyte*.

antigen The type of chemical, carried by a germ, that causes a specific antibody to be produced. This antibody is specialized to destroy the antigen.

appendicular skeleton The part of the skeleton made up of the bones in the shoulders, arms, pelvis and legs.

aqueous humor A liquid which fills the cavity between the eye's lens and cornea.

association neurons Neurons in the brain and spinal cord which interpret information from sensory neurons, then pass instructions to motor neurons.

auditory nerve See *organ of Corti*.

autonomic nervous system The system of nerves which controls involuntary actions.

axial skeleton The part of the skeleton made up of the skull, backbone and ribcage.

axons Nerve fibers which carry information away from a neuron's cell body.

Bile A green liquid, produced by the liver, which breaks up fat into tiny drops so enzymes can break it down.

bone marrow A tissue found inside bones. **Red marrow** is found between the trabeculae of spongy bone, and is where blood cells are made. **Yellow marrow** is a store of fat, found in the marrow cavity of long bones.

Bowman's capsule See *glomerulus*.

brain stem The part of the lower brain leading out into the spinal cord, made up of the **midbrain**, **pons**, and **medulla**. It controls automatic processes, for example. breathing.

bronchi Two thick tubes (each one a **bronchus**) into which the trachea divides, leading into the two lungs. Also the thinner tubes which branch off these first two.

bronchioles Tiny tubes in the lungs, which branch off the thinnest bronchi (**tertiary bronchi**) and end in alveoli.

Capillaries Tiny vessels through whose walls oxygen passes from the blood to the body cells, and carbon dioxide and waste pass in to be carried away. See also *tissue fluid*.

carbohydrates Organic compounds, containing carbon, hydrogen and oxygen. The simple carbohydrate glucose is stored as glycogen in the liver, and broken down during internal respiration to give energy for life.

cardiac muscle tissue A type of striated muscle tissue, made of Y-shaped fibers which interlock to make up the cardiac muscles of the heart.

cerebellum The part of the brain that coordinates muscle movement and balance.

cerebrospinal fluid A thin layer of liquid that cushions the brain in the skull.

cerebrum The largest part of the brain, divided into two joined halves, called **cerebral hemispheres**. It controls most physical, and many mental, activities, as well as the activities of the cerebellum. Its outer layer is called the **cerebral cortex**.

cervix The muscular "neck" of the womb, which opens into the vagina.

chemoreceptors Cells which sense the presence of certain chemicals, such as those on the tongue. See also *olfactory cells*; *taste buds*.

cholesterol A fat-like substance found in animal cell membranes. Too much cholesterol in the body can cause arterial disease.

chromosomes Structures containing the chemical DNA (deoxyribonucleic acid) found in each cell nucleus of a plant or animal. Each chromosome is made up of chemical units called genes, which together contain all the information needed to form that individual organism.

cilia Tiny hairs on surfaces inside the body, such as those on the lining of the human nose, which keep dust and mucus out of the lungs.

ciliary muscles Muscles in the eye which change the shape of the lens.

circular muscles Muscles in the iris which contract in bright light, shrinking the pupil to avoid dazzling.

cochlea A spiral-shaped tube in the inner part of the ear, filled with fluid. It passes sound vibrations to hair cells on the organ of Corti.

co-dominant genes A pair of genes, the instructions of which are both carried out. Neither masks the other.

colon The first part of the large intestine.

cones Cone-shaped receptors in the retina, which are sensitive to red, green or blue light.

conjunctiva A clear layer which covers the cornea and lines the eyelids.

cornea A transparent covering protecting the front of the eye. It is the front, central part of the sclera.

cornified layer The outer layer of the epidermis, made of dead skin cells containing keratin.

corpus callosum A thick band of nerve fibers connecting the cerebral hemispheres.

cortex The outer part of an organ such as a kidney.

cupula A gel-like projection, lying across the inside of the ampulla of each semicircular duct in the ear. It contains hair cells which respond to fluid movement when the head is rotated or tilted, by sending impulses to the brain.

cusps 1. Flaps on the valves between the heart's four chambers, which are forced open and shut, ensuring one-way circulation of blood. 2. Points on certain types of teeth.

cuticle 1. The outer surface of a hair, made of keratin. 2. The pad of thick skin at the base of a nail.

Dendrites Nerve fibers which carry information toward a neuron's cell body.

dentine The substance that makes up the bulk of a tooth, covered by enamel.

deoxyribonucleic acid See *DNA*.

deoxyribose A type of sugar that makes up the "backbone" of the DNA double helix.

dermis The thick, lower layer of skin beneath the epidermis.

diaphragm The flat sheet of muscle under the lungs, which flattens upon breathing in, and relaxes back upward upon breathing out.

diencephalon The central part of the brain, made up of the thalamus and hypothalamus.

digestive glands Glands in the digestive system, which produce **digestive juices**. These contain enzymes that break down food into simpler substances.

digestive tract (or **alimentary canal**) The tube from the mouth to the anus, through which food passes.

DNA (deoxyribonucleic acid) The acid, found in cell nuclei, whose double helix molecules form part of the chromosomes of all living things.

dominant gene The member of a pair of genes which carries an instruction that overrules the instruction carried by the other, **recessive**, gene.

double helix The shape of the DNA molecule, like a twisted rope ladder.

duodenum The first part of the small intestine, where digestive juices break down fats, protein and starch. The resulting mixture travels into the **jejunum**, through whose walls digested food is absorbed. The remaining undigested food passes into the **ileum**, where further absorption and digestion take place.

Ejaculation The squirting of semen out of the penis, caused by the contraction of muscles around the urethra.

electroencephalogram (EEG) A chart used by doctors to record brain wave patterns.

embryo A baby in the first two months of growth in the womb.

enamel The hard, white substance which forms the surface of a tooth's crown.

endocrine glands Glands, such as the adrenal glands, which produce hormones and secrete them directly into the blood. See also *exocrine glands*.

endoscope A flexible tube pushed into a patient's body, used to see and remove things, such as growths.

enzyme A chemical which aids and speeds up internal processes.

epidermis The outermost skin layer.

epididymis A comma-shaped organ lying over the back of each testis, in which sperm are stored.

epiglottis A flap which closes the trachea during swallowing, to prevent choking.

epimysium A tough, protective layer around a muscle.

esophagus (or **gullet**) The tube which carries food from the throat to the stomach.

excretion The removal of waste substances from the body.

exocrine glands Glands such as sweat glands, which secrete substances onto a surface or into a cavity (see also *endocrine glands*).

Fallopian tubes Two tubes, next to each ovary, which carry the ovum into the uterus during ovulation.

fascicles See *muscle fibers*.

fast-twitch fibers Striated muscle fibers which contract quickly, working in short bursts.

fats A group of substances stored in the body as a reserve energy source.

fertilization (or **conception**) The joining of male and female sex cells to form the first cell of a new baby.

fetus An unborn baby after the first two months of its growth in the womb.

fibrin A substance made up of sticky threads, produced by chemical reactions in platelets, which join to form a blood clot.

follicles Deep pits in the skin from which hairs grow.

foreskin A loose fold of skin which protects the sensitive tip of the penis.

Gall bladder A sac in the body which stores bile until it is needed.

gastric juices Acidic liquids made by **gastric glands** in the stomach which break down food and kill germs.

genes The pairs of chemical instructions (each a segment of DNA on a chromosome) which together form the blueprint for building a living thing. See also *dominant gene*.

genetic engineering The extraction of genes from organisms for various uses, for example, in creating new medicines.

genetic modification A specific type of genetic engineering. It involves changing an organism by adding to it the gene for a desired feature from another organism. The changed organism is described as **transgenic**.

genome All the DNA making up an organism.

gingivae The gums.

gingivitis Gum disease, causing bleeding.

glands Organs which produce substances vital to the workings of the body. See also *endocrine, exocrine glands*.

glans The sensitive tip of the penis.

glomerulus A ball of coiled-up capillaries in a nephron, where high pressure filters glucose, water and salts out of the blood into a cup-like structure called the **Bowman's capsule**.

glucagon A hormone which raises the level of glucose in the blood.

glucose The simple carbohydrate broken down in internal respiration to obtain energy.

glycogen A form of starch into which glucose is converted for storage in the liver.

gustatory receptor cells See *taste buds*.

Hair cells Receptors with projecting hairs which are moved, for instance by air or fluid. When this happens, the cells fire off impulses to the brain.

hair erector muscles Tiny muscles in the skin that make hairs stand on end.

hair plexuses Groups of motion-sensitive nerve fiber endings around hair roots.

Haversian canal A channel in compact bone through which blood vessels pass.

hemoglobin A purple-red chemical in red blood cells. It combines with oxygen in the lungs to form oxyhemoglobin.

heterozygous A term describing a person's status with regard to a particular gene pair. If the person has inherited different genes for a characteristic, they are described as being heterozygous for that characteristic.

homeostasis The body's natural regulation of its internal conditions, such as temperature and chemical balance.

homologous chromosomes The pairs into which chromosomes are arranged. Each pair carries gene pairs (one gene on each chromosome) which determine characteristics. Human cells (except sex cells) have 23 pairs.

homozygous A term describing a person's status with regard to a particular gene pair. If the person has inherited identical genes for a characteristic, they are described as being homozygous for that characteristic.

hormones Chemicals made by endocrine glands which help to control the levels of substances in the body.

hypothalamus A part of the diencephalon controlling homeostasis and certain hormones.

Infundibulum A funnel-shaped opening at the end of a Fallopian tube.

insulin A hormone, made in the pancreas, which lowers the level of glucose in the blood.

integumentary system The skin, nails and hair, which together protect the body from damage, infection and water loss.

intercostal muscles Muscles between the ribs which contract on breathing in to expand the chest cavity, and relax again on breathing out.

internal respiration The process by which energy is released from digested food. See also *aerobic, anaerobic respiration*.

iris The colored part of the eye, with muscles which control pupil size.

islets of Langerhans Clusters of cells (each cluster an endocrine gland) in the pancreas that produce insulin and glucagon.

Keratin A waterproof protein of which hair and nails are mostly made.

labia Two folds of skin which surround the openings of the vagina and the urethra.

lachrymal canals Two channels that drain tears from the eye into the nose.

lachrymal glands Two glands, one above each eye, that produce tears to keep the eyes moist and clean, and to fight infection.

lactic acid See *anaerobic respiration*.

lacunae Tiny spaces in bone which house the osteocytes.

lamellae Dense, circular layers of bone which make up compact bone.

larynx (or **voice box**) The organ of speech, found at the top of the trachea.

liver A large organ that produces bile, breaks down amino acids from food into urea, filters some poisons, and stores glycogen, minerals and vitamins.

lymph A liquid containing white blood cells soaked up from tissue fluid.

lymphatic system A network of tubes and organs, containing lymph, which help the body to fight disease.

lymph nodes Small organs that lie in clusters along the lymphatic system which produce white blood cells and trap germs.

lymphocyte A white blood cell that destroys germs by releasing chemicals called antibodies.

macrophages A term used to describe monocytes that are constantly moving around (**wandering macrophages**) and those which remain fixed in an organ such as a lymph node (**fixed macrophages**).

maculae Two gel-like patches, one each in the saccule and utricle of an ear. They contain tiny grains called **otoliths**, which move when the head does, and trigger hair cells to fire off impulses to the brain.

marrow See *bone marrow*.

medulla An inner layer of an organ, such as that of the kidney or brain stem.

meiosis The type of cell division which produces four sex cells, each of whose nuclei contain only half the original number of chromosomes.

Meissner's corpuscles Touch receptors in the skin that send impulses to the brain.

melanin The brown pigment in skin and hair. It absorbs harmful ultraviolet rays.

melanocytes Melanin-producing skin cells.

menopause The time of life, usually between the ages of 40 and 55, when a woman's ovaries stop releasing ova and menstruation stops.

menstruation The shedding of an unfertilized ovum, and the blood-rich inner layer of the womb, which occurs in girls after puberty about every 28 days (each time known as 'having a period').

metabolic rate The rate at which a body converts food into energy.

metabolism The collection of processes in the body involved in producing energy, growth and waste.

midbrain See *brain stem*.

mitosis A type of cell division in which the nuclei of the two resulting cells each have the original number of chromosomes.

monocyte A type of white blood cell that destroys germs by phagocytosis. See also *macrophages*.

motor neurons Neurons which pass instructions from the central nervous system to the body, where they are carried out.

motor-skill memory Memory which recalls how to do actions such as walking.

mucus A slippery liquid made by the mucus membrane, for example the lining of the nose and trachea. It traps dirt and germs, and moistens air breathed in.

muscle fibers (or **myofibers**) Long, rod-shaped cells which, in groups called **fascicles**, make up muscle tissue.

myofibrils Thin cords which join to make up muscle fibers.

myofilaments Thick and thin threads which interlock, making up myofibrils.

nephrons Tiny filtration units in a kidney.

neurons Nerve cells. See also *association, motor, sensory neurons*.

neurotransmitter A chemical released when a nerve impulse reaches a synapse. If enough neurotransmitter builds up, the impulse crosses the synapse to the next neuron.

NREM (non-rapid eye movement) sleep A type of sleep in which the eyes remain still.

nucleotides The chemical units of which DNA is built. Each unit is made up of deoxyribose, a base and a phosphate group.

olfactory cells Chemoreceptors in the nose that send information about chemicals to the brain. The brain then interprets these as smells.

olfactory hairs Tiny hairs that dangle from the roof of the nasal cavity. They are the dendrites of olfactory cells, and absorb airborne chemicals called **odorant molecules**.

organ of Corti A membrane that runs inside the cochlea, containing hair cells. These turn sound vibrations into nerve impulses that are then sent along the **auditory nerve** to the brain, enabling hearing.

ossification The process by which young cartilage skeletons turn into hard bone.

osteocytes Living bone cells.

otoliths See *maculae*.

ova Female sex cells.

ovaries The two female reproductive organs, which produce ova.

ovulation The release of a mature ovum into a Fallopian tube.

oxygen debt See *anaerobic respiration*.

oxyhemoglobin A bright red chemical formed when hemoglobin in red blood cells combines with oxygen. As oxygen is released to the body, the oxyhemoglobin turns back to hemoglobin.

pacinian corpuscles Deep pressure receptors in the skin that send impulses to the brain.

pancreas An organ which produces pancreatic juice and other important substances, such as insulin.

papillae Tiny bumps on the tongue, many of which contain taste buds.

pathogen (or **germ**) A microscopic organism, such as a virus or bacterium, that causes disease.

penis The organ with which a male passes sperm directly into a female's body. It also releases urine.

perimysium A protective layer around a fascicle in a muscle.

periodontal ligament A bundle of tough fibers joining a tooth's roots to the jawbone.

periosteum A thin layer of tissue covering the bones, that contains cells for growth and repair.

peripheral nervous system The network of nerves which carries information to and from the central nervous system.

peristalsis The contractions of muscles in the walls of the digestive tract which push food through it.

phagocytosis The process by which monocytes destroy germs by changing their shape and engulfing them.

pharynx The area at the back of the mouth, where the mouth and nasal cavities meet.

pinna The outer ear flap.

placenta An organ in the womb that provides an unborn baby with food and oxygen from its mother.

plaque A thin, sticky white film that builds up on teeth, formed by bacteria.

plasma A pale yellow liquid which makes up over half the content of blood.

platelets Cell fragments in the blood, with no nuclei, which gather together at an injured area to cause **clotting**. See also *fibrin*.

pons See *brain stem*.

prostate gland A gland surrounding the top of the male urethra which makes some of the fluid in semen.

proteins Natural substances made up of amino acids. They are produced in cells and are essential to growth and tissue repair.

pulp cavity The soft, central area of a tooth.

radial muscles Muscles in the iris which contract in dim light to dilate (expand) the pupil, allowing more light to enter the eye.

receptors The dendrite endings of sensory neurons. They respond to stimuli such as light and heat.

recessive gene See *dominant gene*.

rectum The final part of the large intestine, where the semi-solid waste matter (**feces**) collects until it is released through the anus.

reflex action A type of involuntary action, usually a sudden movement, such as moving the hand away from something hot.

reflex arc The "short cut" route, via the spinal cord only, taken by nerve impulses in a reflex action.

REM (rapid eye movement) sleep The type of sleep where the eyes move, causing a visible fluttering of the eyelids.

renal pelvis The cavity in a kidney where urine collects before it flows into the ureter.

respiratory system The lungs and all the tubes involved in breathing. See also *bronchi; bronchioles; alveoli.*

retina The light-sensitive layer, made of rods and cones, at the back of the eye. It sends nerve impulses to the brain via the **optic nerve**.

rods Rod-shaped receptors in the retina, which are sensitive to light, but not to color.

rugae Folds in the lining of some organs, such as the stomach, which flatten as the organ fills.

saccule A sac at the base of the vestibular system in the inner ear, which is important in sensing head position. See also *maculae.*

sclera The tough, "white" coating of the eye. It is mostly opaque, except for the transparent front (the cornea).

scrotum The sac of skin hanging outside the body of a man, which contains the testes.

sebaceous glands Glands in the skin producing an oil (called **sebum**) that keeps it waterproof and supple.

semen The mixture of sperm and fluids from the seminal vesicles and prostate gland which leaves the penis during ejaculation.

semicircular canals Three loops in the inner ear, positioned on the three different planes of movement. They contain semicircular ducts, and are involved in maintaining balance.

semicircular ducts Channels inside the semi-circular canals. In the swelling at the base of each one (called the **ampulla**) there is a cupula which detects fluid movement.

seminal vesicles Two organs beneath the bladder that produce some of the fluid in semen.

senescence The process of aging.

sensory neurons Neurons ending in receptors. They carry information about stimuli from these receptors to the central nervous system.

sex chromosomes The chromosomes, called X and Y chromosomes, which determine a child's gender. A girl has two X chromosomes; a boy has an X and a Y.

sex-linked genes Genes found only on the X chromosome.

skeletal muscles Voluntary muscles, mostly connected to the skeleton by tendons.

slow-twitch fibers Striated muscle fibers which contract slowly, and can work for a long time without tiring.

smooth muscle tissue A type of muscle tissue, made up of spindle-shaped fibers, which makes up the visceral muscles.

sperm ducts Two tubes which carry **sperm** (male sex cells) from the epididymis into the urethra.

sphincter A ring of muscle, such as that which controls the flow of urine out of the body.

spinal cord A thick bundle of nerves that runs from the brain down channels in the backbone. It carries impulses to and from all parts of the body.

spongy bone A strong, light type of bone made of a mesh of trabeculae with red bone marrow between them.

starch A complex carbohydrate, made and stored for energy by plants. It is broken down to glucose in the body. This is then either stored as glycogen or used for energy.

stereoscopic vision The type of vision in which each eye sends a slightly different view to the brain, which is able to form a 3D image.

stretch receptors Receptors in muscles and tendons which inform the brain about the position of the body.

striated (or striped) muscle tissue The tissue that makes up skeletal muscles.

subcutaneous layer A layer of fat cells beneath the dermis which helps to keep the body warm.

synapse The junction of two neurons.

synovial fluid A lubricating fluid produced in most freely-movable joints (**synovial joints**).

Taste buds Tiny structures in the papillae of the tongue, containing gustatory receptor cells. These chemoreceptors send signals about chemicals in food to the brain, which interprets them as tastes.

tendons Tough bands of tissue which attach muscles to bones.

testes The two male glands which produce sperm.

thalamus Part of the diencephalon that sorts impulses entering the brain, then sends them to other areas for processing.

tissue fluid The fluid between cells which carries oxygen and dissolved food from the capillaries to the cells, and carbon dioxide and waste from the cells to the capillaries.

trabeculae A criss-cross network of branch-like structures that make up spongy bone.

trachea (or **windpipe**) A tube through which air enters and exits the lungs.

transgenic See *genetic modification.*

Umbilical cord A tube, containing arteries and a vein, which connects a baby to the placenta. It is cut at birth.

urea A waste substance produced in the liver from the breakdown of amino acids.

ureter A tube through which urine passes from a kidney to the bladder.

urethra An opening in the body through which urine is released from the bladder.

urine A waste substance produced by the kidneys, made of urea, water and toxic salts.

uterus See *womb.*

utricle A sac between the semicircular canals and the saccule, which is important in sensing head position. See also *maculae.*

Vaccine A dose of a germ too weak to cause disease in a person, but strong enough that they produce antibodies which will resist it in the future. The use of vaccines is called **vaccination**.

vagina The muscular passage which leads from a female's cervix out of her body. It receives the penis during sexual intercourse, and a baby is born through it.

vertebral column (or **spine**) The backbone, made up of 33 interlocking bones called **vertebrae**.

vestibular system A collection of organs in the inner part of the ear which help the body to keep its balance. See also *semicircular canals; utricle; saccule.*

villi Tiny, finger-like projections on the lining of the small intestine which absorb digested food.

viruses Strands of DNA (or a related acid called **RNA**) in a protective coat. They cannot live on their own, but invade living cells in order to reproduce, often causing diseases such as colds.

visceral muscles Muscles in the walls of internal organs such as the intestines.

vitreous humor A stiff, jelly-like substance that fills the central body of the eye behind the lens, keeping its shape.

Volkmann's canals Tiny channels in bone which branch from Haversian canals and carry blood vessels and nerves to the osteocytes.

Womb (or **uterus**) A sac inside the female body in which a baby grows and develops.

X chromosome The type of sex chromosome contained in all ova, and half of all sperm.

Y chromosome The type of sex chromosome contained in half of all sperm.

Zygote The first cell of a new baby, formed when a male cell fertilizes a female cell.

TEST YOURSELF

1. The knee is a type of:
A. ball and socket joint
B. hinge joint
C. gliding joint   *(Page 9)*

2. Which of these terms does not describe the biceps and triceps?
A. an antagonistic pair
B. cardiac muscles
C. skeletal muscles   *(Pages 10-11)*

3. Oxygen-rich blood is carried from the heart around the body via:
A. the aorta
B. the pulmonary artery
C. the pulmonary vein   *(Page 12)*

4. Oxygen is given up to tissue fluid through the walls of:
A. arteries
B. veins
C. capillaries   *(Page 13)*

5. Oxygen is carried around the body in:
A. white blood cells
B. red blood cells
C. platelets   *(Page 13)*

6. The shape of molar teeth makes them particularly suitable for:
A. tearing and stabbing food
B. crushing and grinding food
C. chopping food   *(Page 14)*

7. A tooth starts to hurt when bacteria attack the:
A. enamel
B. dentine
C. pulp cavity   *(Page 15)*

8. The enzyme that is contained in saliva digests:
A. starch
B. protein
C. fat   *(Page 16)*

9. The organ in which bile is produced is the:
A. pancreas
B. small intestine
C. liver   *(Page 17)*

10. Absorption of digested food takes place in the:
A. stomach
B. small intestine
C. colon   *(Page 17)*

11. The main nutrient found in meat is:
A. protein
B. carbohydrate
C. fat   *(Page 18)*

12. Gas exchange takes place in:
A. bronchi
B. bronchioles
C. alveoli   *(Page 20)*

13. Internal respiration that uses oxygen is:
A. anaerobic respiration
B. metabolism
C. aerobic respiration   *(Page 22)*

14. Aerobic respiration is described as:
A. glucose + oxygen + water
   $\longrightarrow$ energy + carbon dioxide
B. glucose + carbon dioxide
   $\longrightarrow$ energy + oxygen + water
C. glucose + oxygen
   $\longrightarrow$ energy + carbon dioxide + water
                                (Page 22)

15. Exercise:
A. allows you to use less oxygen
B. increases the pulse rate
C. keeps the heart muscles beating at a steady rate   *(Page 23)*

16. Chemical compounds controlling the level of substances in the body are:
A. glomeruli
B. hormones
C. nephrons   *(Page 24)*

17. Receptors are the sensitive nerve endings of:
A. motor neurons
B. association neurons
C. sensory neurons   *(Page 26)*

18. The largest part of the brain is:
A. the cerebrum
B. the cerebellum
C. the brain stem   *(Page 28)*

19. Pacinian corpuscles are found in:
A. the blood
B. the brain
C. the skin   *(Page 30)*

20. The size of the pupil of the eye is controlled by:
A. changing the shape of the lens
B. the iris
C. rods and cones   *(Pages 32-33)*

21. A structure called the oval window is found in:
A. the ear
B. the eye
C. the kidney   *(Page 34)*

22. Taste buds that detect bitter tastes are thought to be at the:
A. front of the tongue
B. back of the tongue
C. sides of the tongue   *(Page 37)*

23. Conception is:
A. fertilization of an egg by a sperm
B. another word for sexual intercourse
C. ejaculation   *(Page 39)*

24. The organ that provides a baby with food and oxygen from the mother is:
A. the placenta
B. the uterus
C. the vagina   *(Pages 38-39)*

25. A woman's periods happen:
A. when the uterus lining is lost
B. when ovulation takes place
C. when the uterus lining starts to thicken   *(Page 41)*

26. The form of cell division that produces sperm is:
A. fertilization
B. meiosis
C. mitosis   *(Page 42)*

27. Which of the following does a measles vaccine contain?
A. a weak dose of the measles germ
B. antibodies to the measles germ
C. lymphocytes   *(Page 49)*

28. Antibiotics are effective against:
A. some viruses
B. some bacteria
C. all germs   *(Page 51)*

29. Which of the following is not an alternative treatment?
A. laser surgery
B. homeopathy
C. acupuncture   *(Pages 51-52)*

30. Which of these an illegal drug?
A. nicotine
B. alcohol
C. heroin   *(Page 53)*

60 *Answers*

1.B 2.B 3.A 4.C 5.B 6.B 7.C 8.A 9.C 10.B 11.A 12.C 13.C 14.C 15.B 16.B 17.C 18.A 19.C 20.B 21.A 22.B 23.A 24.A 25.A 26.B 27.A 28.B 29.A 30.C

INDEX

You will find the main explanations of terms in the index on the pages shown in bold type. It may be useful to look at the other pages for further information.

ACKNOWLEDGEMENTS

PHOTO CREDITS
(t = top, m = middle, b = bottom, l = left, r = right)

Corbis: 1 Charles O'Rear; 36 (b) Hulton-Deutsch Collection; 38 (t) Michael Freeman; 41 (br) Jenny Woodcock, Reflections Photo Library; 42 (tr) Bettmann, (mr) Bettmann, (br) Corbis; 51 (br) Science Pictures Limited; 52-53 (main) Bill Ross.
Digital Imagery © Copyright 2001 PhotoDisc, Inc: 6-7
© Digital Vision: 24-25; 55; **Images Colour Library**: 5
Science Photo Library: **cover** Hugh Turvey, Dr Gopal Murti; 8-9 (t) Dr Yorgos Nikas; 10-11 (main) Manfred Kage; 12 (r) John Daugherty; 14-15 (t) National Cancer Institute; 16-17 (t) Tek Image; 22-23 (t) Alfred Pasieka; 26 (b) Biophoto Associates; 28-29 CNRI; 30 Mehau Kulyk; 32-33 (b) Prof. P. Motta/Dept of Anatomy/University "La Sapienza", Rome; 34-35 (b) David Parker; 40 (t) Dr G. Moscoso; 44-45 Biophoto Associates; 46 (tr) A. Barrington Brown; 47 (tm) Simon Fraser; 48-49 (t) Juergen Berger, Max-Planck Institute; 50 (t) Mehau Kulyk, (b) Hugh Turvey; 52 (bl) BSIP Boucharlat.
Telegraph Colour Library: 2 Simon Bottomley
Opera North/Richard Moran: 23 (mr) Roderick Williams as Figaro in Opera North's production of *The Barber of Seville*.

ILLUSTRATORS
Simone Abel, Sophie Allington, Rex Archer, Paul Bambrick, Jeremy Banks, Andrew Beckett, Joyce Bee, Stephen Bennett, Roland Berry, Gary Bines, Isabel Bowring, Trevor Boyer, John Brettoner, Peter Bull, Hilary Burn, Andy Burton, Terry Callcut, Kuo Kang Chen, Stephen Conlin, Sydney Cornfield, Dan Courtney, Steve Cross, Gordon Davies, Peter Dennis, Richard Draper, Brin Edwards, Sandra Fernandez, Denise Finney, John Francis, Mark Franklin, Nigel Frey, Giacinto Gaudenzi, Peter Geissler, Nick Gibbard, William Giles, David Goldston, Peter Goodwin, Jeremy Gower, Teri Gower, Terry Hadler, Bob Hersey, Nicholas Hewetson, Christine Howes, Inklink Firenze, Ian Jackson, Karen Johnson, Richard Johnson, Elaine Keenan, Aziz Khan, Stephen Kirk, Richard Lewington, Brian Lewis, Jason Lewis, Steve Lings, Rachel Lockwood, Kevin Lyles, Chris Lyon, Kevin Maddison, Janos Marffy, Andy Martin, Josephine Martin, Peter Massey, Rob McCaig, Joseph McEwan, David McGrail, Malcolm McGregor, Christina McInerney, Caroline McLean, Dee McLean, Annabel Milne, Sean Milne, Robert Morton, Louise Nevet, Martin Newton, Louise Nixon, Steve Page, Justine Peek, Mick Posen, Russell Punter, Barry Raynor, Mark Roberts, Andrew Robinson, Michael Roffe, Michelle Ross, Michael Saunders, John Scorey, John Shackell, Chris Shields, David Slinn, Guy Smith, Peter Stebbing, Robert Walster, Craig Warwick, Ross Watton, Phil Weare, Hans Wiborg-Jenssen, Sean Wilkinson, Ann Winterbottom, Gerald Wood, David Wright.

Every effort has been made to trace the copyright holders of the material in this book. If any rights have been omitted, the publishers offer to rectify this in any future edition, following notification.

American editor: Carrie A. Seay. With thanks to US expert Dr Janet Mercer.